# Growing Gladioli

**E.W. Anderton and R. Park**

# *Growing Gladioli*

'. . . the African Gladioli
will become great favourites
when their beauty, facility of
culture, and the endless variety
that may be produced from seed by
blending the several species
is fully known.'

*William Dean Herbert, 1820,*
*(Hort. Society, London)*

**Christopher Helm**
London

**Timber Press**
Portland, Oregon

© 1989 E.W. Anderton and R. Park

*Line illustrations by* Eric Anderton

Christopher Helm (Publishers) Ltd
Imperial House, 21-25 North Street
Bromley, Kent BR1 1SD

ISBN 0-7470-2621-1

A CIP catalogue record for this book
is available from the British Library

First published in North America in 1989 by
Timber Press, Inc.
9999 S.W. Wilshire, Portland,
Oregon 97225

ISBN 0-88192-165-3

Typeset by Leaper & Gard Ltd, Bristol, England
Printed and bound in Great Britain by
Billing and Sons Ltd, Worcester

# Contents

# Colour Plates

1   'Vicki Lin'. A large-flowered exhibition cultivar.

2   'Parade'. A prize large exhibition gladiolus.

3   'Moon Mirage'. A frequent Grand Champion from the Giant size exhibition class.

4   'Moon Mist'. An example of the ruffled floret.

5   'Jester'. An exotic sport with laciniated petals.

6   *Gladiolus nanus*: a spring-flowering species hybrid, 'Amanda Mahy'.

7   A prize-winning selection of florets of primulinus hybrids.

8   A champion primulinus hybrid seedling. 'Rutherford'.

9   'Amy Beth', a show miniature.

10  'Lynn', a red show miniature.

11  'Winsome', a formal show miniature.

12  'Pegasus', a small-flowered show gladiolus.

13  'Tendresse', a show gladiolus in the medium size.

14  A prize-winning trio: 'Carrara', 'Moon Mirage' and 'Grand Finale'.

15  A championship basket display.

16  A top-rated Canadian cultivar, 'Drama'.

# Figures

# Acknowledgements

No one person, or even two, could from their own personal experience have acquired or retained all the information contained in this book. The most important acknowledgement therefore is to the numerous gladiolus friends throughout the world whose contributions to our fund of knowledge have made this book possible.

Many have made their contributions inadvertently by staging superb exhibition spikes and telling how such perfection was achieved, by explaining over a friendly cup of tea how they overcame some problem, or by just giving friendly comment or constructive advice on some apparently futuristic idea.

To these friends must be added those growers, hybridists and academics who over the years have kept up an on-going correspondence contact, and contributed knowledgeable articles to the *Gladiolus Annual* and the North American Gladiolus Council bulletins. To all who have contributed in any way, our grateful thanks.

More specifically we wish to thank those friends who have made more direct contributions to the book by supplying archive material, or allowing us to use 'as yet unpublished material' which they hoped to use one day, and 'never got round to it!'.

For information on the species, we are indebted to Tony Hamilton and Lee Fairchild, and to Mrs W. Howarth for the loan of Bill Howarth's records.

To Carl Fischer, Len Butt and Alex Summerville for their many kindnesses in supplying information, photos and corms of their hybrids.

To Reg and Marlene Powyss-Lybbe for giving us full access to their 'Grand Champion Production Line' in Alberta, and revealing all their secrets.

To Dr Bob Magie, and Professor Gary Wilfret of Florida for their hospitality, and for providing information on the Florida gladiolus scene, as well as introducing us to the gladiolus wonderland of Mr Preston's farm at Manatee Fruit Co.

Thanks to John Evans for his information on feeds and fertilisers, and to two charming ladies for their historical research. Miss Angela Evans sought out details of the life of William (Dean) Herbert, and Miss Lugtendorp of Leiden in the Netherlands did 'on the spot' research on the Linneaus family. Our sincerest thanks to these three.

Thank you Ingineer Igor Adamovic for your multi-lingual co-operation and information on the gladiolus in East Europe, and in particular for divulging information on the fragrant hybrids. The patience and understanding during our 15-year friendship and for answering hundreds of questions posed by a 'pedestrian old man' trying to keep up with a modern whizz-kid, is much appreciated.

To all those generous friends who loaned us illustrations, photos, slides, etc. to use in our book, our appreciation and thanks.

Finally, though of course not in any way less sincere, our thanks to those who guided us through the minefield of the mechanics of making a book: The editors, publishers and printers, the typists (now WP operators) and those who physically produced the book. We read their minds — 'They may know a lot about gladioli, but what they know about book making is. . . .'

*Eric Anderton*
*Ron Park*

# Introduction

It is probable that no genus of plants has contributed more to world horticulture than the genus *Gladiolus*. Although it is the species from Southern Africa that have made the most spectacular contribution, it is the Eurasian species that were first recorded. Several years ago in post-war Paris, I acquired a painting in silk of a wild gladiolus. It seemed to me at the time to be a stylised illustration of a species similar to what was then called *Gladiolus byzantinus*. A reliable art expert said the picture was at least 500 years old and of Chinese origin. This is, in my long experience of the flower, the oldest illustration of a gladiolus I have ever seen.

There is a legend about the gladiolus recorded in ancient Greek, and the Roman soldier and writer Pliny refers to what is almost certainly *Gladiolus segetum* of the cornfields in his writings, where he records the use of small dried gladiolus corms as amulets. He writes also of the corms being roasted and eaten like chestnuts.

The old Greek legend tells of two brothers who fell in love with the same girl and neither one would concede to the other. Eventually brotherly love gave way to intense hatred which mounted until it reached a point where they vowed to fight each other to the death, but both brothers were fatally wounded in the battle. In the final moments of their lives they plunged their swords in to the ground and, so the legend says, from that very spot grew the first two gladiolus plants with leaves like little swords and blood-red flowers marked in their hearts with water-white splashes from the girl's tears.

It is a charming legend and well worth perpetuating. Sadly, the first botanist who reads the legend will doubt if there are any gladiolus species endemic to Greece that answer to the legend's colour description — but then taxonomists are known as horticulture's spoil-sports.

In the Middle Ages, when herbals, potions and salves were the health-care specifics of the times, many illustrations of *G. segetum* appeared in the journals of the herbalists of those days. The native British gladiolus species of that era was almost certainly *G. segetum* since most references are to the 'Corne Flagge', its purple flowers and small, flat round corms (bulbs), and to its being dug up at reaping time by serfs at harvest. Herbalists claimed that the roasted corms were good for relieving stomach pains and 'colic' and were very much in demand.

The gladiolus species under its varied names and descriptions continued to be written about by the botanists and plantsmen of Britain, France and Holland throughout the sixteenth and seventeenth centuries. Until about 1680 all references to gladiolus were confined to the species from European, Mediterranean or Asia

Minor habitats. Interest in the gladiolus species as garden subjects began to diminish around 1670–80, when rather than as a horticultural item the gladiolus had become the symbol of botanical argument and personal academic prestige. After all, the centre of the arguments was the precise colour of a small number of dull purple and rather boring flowers.

However, the enthusiasm was rekindled about 1690 following the establishment in South Africa of the Dutch East Indies Company. This company set up a trading post with a ship refuelling and revictualling port near Cape Town, and the mainly Dutch staff established a garden too. The garden was initially stocked by the use of seeds, bulbs and plants brought out by ship from Rotterdam. Gradually the European plants were augmented by native South African plants gathered by employees and seamen on their forays into the interior.

Obviously the collectors had been attracted by the native Cape species of gladioli, for in about 1690 a shipment of seeds, plants, bulbs and corms arrived in Rotterdam, accompanied by paintings and sketches in colours of the plants in bloom. There must have been a number of gladiolus species in the consignment because in 1692 a Dr Plunkett was able to give descriptions and botanical details of a number of them including *G. alatus*. The colour range of the flowers Dr Plunkett described, together with news that some were fragrant, rekindled interest in the species *Gladiolus*, and from then on the Eurasian species were very much passé as garden material. It was Cape species forward.

The great Swedish botanist Linnaeus, who invented the botanical binomial system, described *G. angustus* in 1737 after he had seen it in a friend's garden — and incidentally had travelled many miles to see it. By 1753 Linnaeus had established the genus *Gladiolus*, and by the time he died in 1778 he had classified and described at least 30 species (or subspecies). The original documents establishing the genus can be seen in the archives library and museum at Leiden — the Linnaeushof Museum in Holland.

Many of the South African species were grown in botanical and private gardens — gladiolus species collecting had become the 'in' thing in botanical circles. Early in the 19th century inter-species hybridisation was well under way, and amongst those involved was the famous plantsman William (Dean) Herbert. Addressing the Horticultural Society in London in 1820, Dean Herbert predicted that the African gladiolus species would become great favourites of florists and gardeners because of their grace, beauty and ease of culture. He also foresaw that the range of colour in the Cape species of South Africa would place in the hands of hybridists the key to an infinite variety of colour and fragrance.

The famous Dean lived long enough to see some of the hybrids between the South Africa species as well as some of the forerunners of our modern summer-flowering gladioli. He could hardly have envisaged the scale, scope and distribution of summer-flowering gladioli that the late twentieth century has brought. Were he to return today he might be a little disappointed that his favourite fragrant species *G. tristis* had not been the progenitor of modern fragrant giants of good garden quality. He might also have been a trifle surprised that the first genuinely fragrant cross between a garden gladiolus and the Abyssinian *G. callianthus* (*Acidanthera bicolor* of former days) had not been achieved until 1973 — and then in New Zealand.

He would undoubtedly have been really surprised and pleased to learn that the gladiolus had given rise to a multi-million-dollar industry in the USA and Canada, growing over 2,000 different cultivars running into several million corms annually. Were he now to see the endless acres (hectares) of commercial gladioli in Florida and California in full bloom in November and December he would have realised to the full the validity of those predictions of 170 years ago. And he would be thrilled and proud to learn of the magnificent efforts of a few Dutchmen in the late 1940s as they struggled to revive a gladiolus industry decimated by war-time damage. These men had resurrected the gladiolus in time of crisis just as their predecessors of the Dutch East Indies Company had given the gladiolus the 'kiss of life' by discovering and introducing the South African species.

Everything that we have today that makes the gladiolus a favourite flower the world over is the result of the combined contributions of knowledge, expertise, and technology from many people in many places during the 170 years since the Dean's vision of the gladiolus cornucopia was revealed to the public.

The aim of this book is not to 'out-statistic' previous books on gladioli, nor to create or enter into any controversy about the taxonomy and nomenclature of the 200 or so gladiolus species that are believed to exist. It is to try to pass on our enthusiasm and love of the 'friendship flower' to those who do or will share our feelings. We hope it will fire an interest in some of the modern types of gladioli now available, or create a desire to cultivate some of the now endangered species.

Above all, this book aims to help the reader to get the best from his gladioli whichever type he chooses to grow, and to create an entreé to the ever widening spectrum of colour, and of beauty, of our favourite flower.

# *Chapter 1*  Gladiolus in the Wild

This book makes no claim to being a scientific treatise on the species *Gladiolus*, nor does it attempt to clarify or verify the various changes in nomenclature that some professional botanists have made in recent years. The gladiolus species we shall describe and bring to your attention are those which have either played a part in the development of the modern garden hybrid gladioli, or are of themselves worthwhile and interesting garden subjects or greenhouse plants.

Plants of the genera *Gladiolus* and *Homoglossum* belong to a closely related group called the Gladiolinae. This family is rich in garden species such as crocus and iris in the northern hemisphere, and freesia, ixia and watsonia in the southern hemisphere. The botanical differences between *Gladiolus*, *Homoglossum* and *Acidanthera* are small (if not negligible in the eyes of the gardener and plantsman), so that we hope we will be forgiven when we use the term 'gladiolus' to include plants which are, in strictly scientific botanical terms, just not 'glads'. Botanists have not even now finalised the number of gladiolus species known to exist and to have been scientifically verified. The likely total is between 180 and 200 and new ones still come to light, though much less frequently than 20 years ago.

The distribution of the species is very wide throughout the temperate zones of the world. Excluding the arid zones of the Near and Middle East and Africa, gladiolus species are found throughout Africa, southern Europe, Asia Minor, the Middle East, Turkey and Iran. For easy reference a map is included (Fig. 1) and this gives the general area of distribution of the various groups of species. Most of the known species occur in southern Africa and of these the majority come from Cape Province in South Africa. It is within the group of species from the winter rainfall areas of South Africa one finds the species that are fragrant and with the greatest diversity of form, shape and colour.

## Protection and Conservation

Many of the gladiolus species which we describe in this book are threatened with extinction. Fortunately a number are either being grown commercially or are being cultivated by enthusiasts and botanic gardens. Other species are saved only from extinction by their habitat high in the mountains in 'far away places with strange sounding names', or in more recent years, in areas where for military or political reasons, even 'botanists fear to tread'. Most of the Eurasian species are not endangered as yet; sadly the situation in Africa, in particular South Africa, is quite different.

Gladioli are not the only threatened plants, and sad as it may seem it is their beauty and rarity that threatens them. Despite all the conservation laws, the decline in numbers of many species continues, and thoughtless picking of flowers has severely decimated many colonies of wild gladioli. The best defence we have for the sole British wild gladiolus species is that its habitat and location is known only to a few dedicated conservationists who will protect it to the end. If a flower is picked from a tree or shrub it does little harm. Unfortunately, when a gladiolus flower is picked the timing of the act and the amount of leaf taken with the flower stem virtually assures the death of the corm and cormlet, as well as robbing the plant of any possible seed.

The picking of wild gladioli is not sufficiently widespread as to be the major cause of the loss of the species in South Africa. We believe that two other factors have much more significance. Urban, industrial and agricultural expansion have destroyed the habitat of many of the rarer gladioli, and there is not enough interest or incentive for preservation projects to restore, transplant or replace species gladioli in South Africa. Any species gladioli that are not removed to safety before the habitat is destroyed, for whatever reason, will vanish unless someone cultivates new plants. Apart from national institutions, there is little interest among the general public in South Africa for cultivating replacement stock of species gladioli. At present the cultivation of gladiolus species by commercial nurseries is very limited.

Great credit must go to certain garden associations in South Africa, and to the National Botanic Gardens (RSA) at Kirstenboch and the Caledon Municipality for their endeavours via the wild flower nursery and the Caledon Wild Flower Fair. These organisations have kept a seed bank of many threatened wild flowers of the area, including several rare and endangered species of gladioli. Serious-minded conservationists are welcomed by these organisations and seed of wild gladioli are obtainable at very low cost. South Africans, in fact, are very keen to contact overseas gardeners who will devote time and patience to reviving or preserving their native flowers. Leisure gardeners in South Africa and commercial farmers are pre-occupied with local problems far more pertinent than gladioli species and are willing to delegate preservation to interested gardeners.

May we hope that you, dear reader, will be sufficiently enthused about wild (species) gladioli to write for a few seeds and play a minor part in securing a place in the future for some of these delicate little beauties, remembering that nearly 200 years ago these plants triggered off a wave of hybridising that gave us the modern garden gladiolus we now take for granted. We can assure you that the thrill of seeing the first wild gladiolus you have ever raised from seed is unforgettable.

## South Africa

The genus *Gladiolus* is remarkable if not unique amongst the Iridacae (Iris) family in South Africa in that there is not a weekend in the whole calendar year when there is not a verified species of

*Gladiolus* in bloom somewhere within a 250-mile (400km) journey from Cape Town.

It is possible, with the help of a book of the flora of South Africa, to find a different species in bloom each weekend from January to December. The recorded species and variants are remarkably constant in their flowering dates and are to be found flowering in the location of their original discovery decade after decade, although from time to time the dates may be retarded or advanced by climatic conditions. After particularly severe veldt fires in localised areas some species may even miss a year or so but, provided the colony includes a few mature corms that have pulled themselves deep enough, the corms will quickly regenerate from side shoots or cormlets at this lower depth.

In 1774 a young Swedish botanist, Sparrman, found the earliest form of *G. gracilis* in bloom at the end of May in the hills above Simonstown. His description records and specimens are in the Linneaus Herbarium in Leiden. In June 1930 Professor Barnard, guided by Sparrman's information, found it again near to where Sparrman first collected it. Dr Joyce Lewis visited the same location in 1962 with Professor Barnard and that particular form of *G. gracilis* was still there.

Different forms of species (usually colour variants) may be found in flower in different areas from May to September, but each local population keeps very close to its recorded flowering time. By a painstaking search of herbarium records it is possible to make a catalogue of dates and locations which will give a flowering programme for the whole year. Though this may sound a not too complicated task it would involve journeys of thousands of miles in total, in weather conditions from heat to chilling rain, and from high mountain to river valley and sandy plains.

South African species have contributed uniquely to the hybridisation that has given us the modern garden gladiolus hybrid, for not only have these species contributed colour and form but by virtue of the wide variation of habitat have made the garden hybrids a very adaptable and healthy subject capable of growing in locations far removed from their origins. South African species may be found growing in mountainous areas by clear water rivers and falls, on the edges of arid heat desert in Namaqualand and perhaps commonly on the edges of cultivated lands. A claim is made in South Africa that it is possible to find more different species of gladioli on a 160-acre (64 hectare) farm in Little Drakenstein that can be found in the whole of Eurasia.

It is perhaps superfluous to say that irrespective of location all species of gladiolus are botanically the same, that is mechanically speaking. Genetically there is a very wide variation in chromosome number, and this directly affects the fertility of species should interspecies hybridisation be involved. Typically all gladiolus species have a thickened underground stem known as a corm. Flowers are invariably bi-sexual, that means each flower carries both male and female organs within the flower-tube (perianth). Normally for fertilisation to occur the pollen from one plant must land on the stigma of another plant — this is cross-pollination. Few species give high seed yields if left to self-pollination, and an isolated gladiolus species plant grown, for example, in a pot would require to be artificially pollinated.

In nature cross-pollination is effected by insects, only occasionally by wind and some are even pollinated only by specific insects. If the specific insect is a night-flying moth then the plant will have a fragrance in the evening or at night to attract the insect. Others have sweet nectar cells at the base of the tube to attract day-time insects to the nectar — and pollinating the flower '*en passant*'. The species that are normally wind-pollinated have stamens and stigma that protrude beyond the perianth, and the seeds have large wings.

None of the known *Gladiolus* species are indigenous to Canada, the United States, South America or Australasia, and where gladioli have been found in the wild in these countries they have proven to be garden escapes of earlier species or species hybrid origins. As examples of this some variants of the former *G. psittacinus* — now *G. natalensis* — have been found in California, Oregon and British Columbia. Similarly, in Western Australia, Tasmania and North Island, New Zealand, garden-escape forms of *G. byzantinus*, *G. natalensis* and the older *G. nanus* hybrids have been discovered growing in uncultivated tracts of urban or near-urban land.

## Propagation

All species of gladiolus can be propagated sexually, by pollination and setting seed, and the subsequent raising of plants from the seed. True species are capable of producing seeds if pollinated by compatible pollen of the same species. Increasing stock can also be done using cormlets, which will give plants identical in every respect to the parent plant. In some cases a very large number of cormlets are attached to the mature mother corm. The Cape species normally produce over 50 cormlets per corm in a season. In the case of the smaller-flowered species (e.g. *G. tristis*, *G. carinatus*) the cormlets are quite tiny and must be very carefully handled. Cormlets will produce full-size corms in a growing season provided they are prevented from flowering for the first year.

Species and subspecies of the *G. natalensis* group produce many cormlets in clusters on the mature corm. The larger cormlets are best dealt with separately and in a different way to the smaller cormlets. Large cormlets should be dehusked (the outer husk peeled off), soaked overnight in clear water and planted singly in peat pots of about 3½in (8cm) dia. and at least 2in (4.8cm) deep. Smaller cormlets need not be dehusked but should be soaked and planted in peat and soil mixture at a depth of 2in (4.5cm).

It is very unfortunate however that some of the most beautiful and most desirable species are not productive of cormlets and seldom divide or produce multiple corms. This failing in one of the most beautiful of species — the Caledon Bluebell (*G. bullatus*) — has now led to its being a threatened species.

## Habitats

With very few exceptions, the wild species of gladioli do not grow continuously but grow, flower, self-seed and then enter into a

9

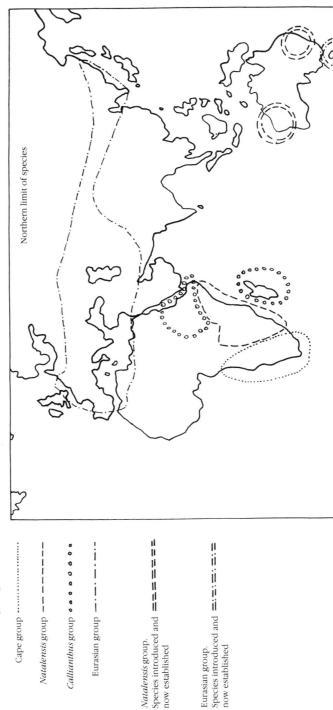

*Figure 1* *The distribution of the genus* Gladiolus

Cape group ⋯⋯⋯⋯⋯

*Natalensis* group – – – – – –

*Callianthus* group ∘ ∘ ∘ ∘ ∘ ∘ ∘

Eurasian group – ∙ – ∙ – ∙ –

*Natalensis* group.
Species introduced and ≣ ≣ ≣ ≣ ≣
now established

Eurasian group.
Species introduced and ≝ ≝ ≝ ≝ ≝
now established

Northern limit of species

Southern limit of species

prolonged dormancy period. The above-ground plant develops and dies back over a period of from 80 to 140 days, the length of growing actively being largely determined by the climate conditions. The main factors influencing the length of the growing period of the species are number of daylight hours, soil temperature, rainfall and air temperature. Of all the species only the Eurasians have any frost resistance.

The South African species from the Cape Province area are triggered off from dormancy by the onset of the late autumn rains (in March), the leaf starting to emerge within a few days of the first rain. After only a few weeks the flower stem appears and the buds show colour quite quickly (in *G. tristis*, for example) thereafter. The stem has usually four or five florets, each floret lasting only about two to three evenings. Once pollinated by night-flying insects the floret will quickly fade. As the temperature increases the seed pods develop very rapidly, and by the end of the rainfall season the seed pods turn brown and open up to scatter the seed in the gusty winds. The whole process is over in about 60 days. The habitat of most of this group of species is in light sandy to gritty soil, the corms at a depth of 3 to 4 in (7 to 10 cm). The soil dries out quickly and is very well drained so that corm harvesting is quite easy. The ripened corms are ready when the soil is warm and dry, so that the procedure usually employed for lifting and storing garden gladioli are quite unnecessary.

The summer-flowering species — our old primulinus friends — have climatic conditions similar to the Mediterranean area, again with alluvial (river bed) soil, with sand and grit if near the rivers. The cycle of shoot-to-seed is longer than that of the Cape species — about 60–65 days to flowering from emergence of the leaf shoot, with a further 25–30 days for seed pods to set. This group has larger florets and larger flower-heads (15–21 buds) thus is much larger in flower. Corms are ripe about 30 days after the first floret appears. A long day length and pleasantly warm temperature throughout the autumn and winter months causes the plants to stay green for longer than is usual for European-grown (or North American) garden hybrids. It is inadvisable to wait for the leaves to turn yellow before lifting the corms of these species, Bitter experience has taught us that the true species in this group have yellow leaves only when something is wrong.

One of the species in this group — formerly known as *G. psittacinus* v. *cooperii* — was brought from South Africa to the Netherlands for use in a breeding programme in 1949–50. This particular species was a very late-flowering variant, very tall with rain-resistant leaves, and leathery, hooded florets of robust substance. As part of the project to restore the Dutch gladiolus industry, the *cooperii* hybrids were used extensively. One hybrid which flowered as late as the end of September, even into October, was a commercially successful tall, silver-pink cut flower, 'Elan'. It propagated remarkably well by virtue of its production of very large cormlets ³⁄₈ in (1 cm) in diameter. The cormlets flowered first season from a March planting. The *cooperii* hybrids found the sand dune soils of Noordwijk very much like home ground, and to the delight of the Dutch growers they flowered so late that the thrips had hibernated before the flowers were there to be eaten.

The Eurasian species have no one habitat, spread as they are over such a wide area. We have species of this group at near sea level in sandy soil, others 9,900 ft (3,000 m) above sea level in mountain areas of poor soil, in cultivated areas from 500 ft (170 m) to 1,000 ft (340 m), and even one in salt marshes (*G. palustris*). It is also from mountain habitats that the now *G. callianthus* originates, in Ethiopia. The multiplicity of species involved in the development of the garden hybrid gladioli readily explains their adaptability to soil conditions – the one constraint factor for species is the pH level of the soil. Gladioli seem to do best at pH 6.5–7.5 i.e. acid to neutral.

Species that are wind-pollinated are mainly Eurasian, Ethiopian, or from what is now Zimbabwe, where there are dry day-time breezes to move about both pollen and ripe seed. Amongst the Eurasian species, a quite common one, *G. segetum*, is to a large extent self-pollinating.

## Distribution

For ease of reference the species, to which we will refer in a later chapter, have been divided into four separate groups on a purely geographical basis. Certain species will however be found on the periphery of two areas, as the botanical records show these, even if as a variant or subspecies, to have been found and recorded in the different locations.

The first and largest group are the winter-growing gladioli of South Africa from the winter rainfall areas of Cape Province, roughly in a triangle, formed with Cape Town on one corner, with the Springbok on Orange River another, and almost due east of Cape Town at Mossel Bay or Port Elizabeth the third.

The second group are the summer-flowering gladioli of southern and central Africa, still better known under the old name of *G. primulinus* and its variants or subspecies, now collected under the general title *G. natalensis*.

Thirdly, both numerically and in botanical importance comes the Eurasian group, which are distributed throughout southern and western Europe, the Mediterranean, including North Africa, Turkey and Iran. *G. illyricus* and *G. byzantinus* are typical of this group.

As the fourth group we have separated out the species from East Africa and Madagascar. This grouping allows us to include *G. callianthus* and allied species of what was formerly *Acidanthera*. *G. callianthus* is of much greater interest and importance now since it has been used extensively in the production of fragrant hybrids by crossing with summer-flowering garden cultivars. The resultant hybrids described are known as Callianthea or Gladanthera.

# *Chapter 2*  The History of the Gladiolus

One of the greatest difficulties to be overcome in writing the history of the gladiolus is to know where to begin. We know of learned botanists who firmly believe that the earliest references to gladiolus are in the Bible, and these gentlemen opine that 'consider the lilies of the field — they toil not neither do they spin, yet I say unto you that even Solomon in all his glory was not arrayed like these', is a reference to *Gladiolus segetum*, the Corn Lily. Archaeologists have carefully examined ninth and tenth century etchings, stone carvings and old eastern silk paintings where flowers are depicted, and have shown that amongst the flowers are forms which resemble to a very high degree the present-day Eurasian gladiolus species *G. communis*.

It is not easy in the late twentieth century, overwhelmed as we are with the image-making wonders of the electronic age, to imagine how people managed to record accurately the colour and form of flowers without a colour camera, or how they could describe colour in the days before graded colour charts. The colours they could produce in their drawings were limited to the colourants and pigments available. Later, when glass technology improved, we had some better colour reproductions of flowers, but all evidence is that before 1700 AD all the gladiolus-like flowers of which we have any kind of record as being of European or Asian origin were a sort of dull lavender–purple, consistent with *G. segetum* and *G. illyricus*. The water-colour paintings of the period 1685–1740, called 'codices' (sing.: codex), by various artists, are of little botanical value, and from these paintings it is virtually impossible to identify or classify with certainty any of the true species.

The earliest record we have of non-European/Asian gladioli being grown in Europe is around 1685–6. It is believed that water-colour paintings in the Codex Simon van de Stel are of South African species that were found by the Van der Stel expedition to Namaqualand at that time. It is not possible now to determine whether the water colours were from sketches done *in habitat* or if plants were brought back and subsequently painted from life. In 1727 Boerhave wrote an account of the gardens at Leiden in Holland in which he mentions the presence of a South African species *G. angustus*. Later, when the Swedish botantist Linnaeus reached Leiden to begin his work on the Latinised binomial system of plant classification, he found this and other species growing in Holland and England. Linnaeus recorded six distinct and stable species at that time and these included *G. tristis*, *G. communis*, *G. segetum* and *G. angustus*. It is a matter of record that *G. tristis* was growing in the Chelsea Gardens in 1745.

The next highlight in the botanical history of the gladiolus species was 1760 when Burman senior sent to his son in Uppsala a large chest containing plants, bulbs, seeds and a whole series of botanical notes, sketches and colour paintings. As a result of this collection, and by four months continuous study of the specimens by Linnaeus, Schreber and Burman junior, 75 new South African species of plants were identified and registered. Many new gladiolus species were amongst the new registrations, including *G. alatus*.

Between the years 1760 and 1784, when the Linnaeus father and son were the activists in describing, registering and preserving the gladiolus species that arrived from numerous sources year by year, many confusing names were given to species from South Africa and elsewhere. The year 1784 can be considered as the end of the Linnaen period of classification of gladiolus species, for by the end of it both Linnaeus father and son were dead, the Linnaen collections had been sold to Smith, and finally Murray had published the 14th edition of *Systema Vegetabilum*, incorporating all the recognised species from the Linnaeus records.

Between 1784 and 1790 many botanical expeditions, both commercial and private, were undertaken to South Africa and to Cape Province in particular. Plants, seeds, descriptions and sketches of many South African gladiolus species were listed in the famous publication *Hortus Kewensis* emanating from Kew Gardens, where indeed many of the new species were from seed and verified. The first of the famous *Curtis Botanical Magazines* was published in London in 1790, and in that edition in October 1790 the first illustration of *G. cardinalis* appeared. This is a very important factor in the history of the gladiolus; it is the start of the pedigree chart of garden hybrids and an indisputable record that *G. cardinalis* was one of the parent species. Kew Gardens raised a significant number of corms of *G. cardinalis* between 1790 and 1795, and distributed them widely to the gardeners of the era.

One of the great gardeners of the period was William (Dean) Herbert, who hybridised prolifically from his collection of gladiolus species said to be of 32 species and subspecies. By the year 1820 Dean Herbert had raised and generously distributed to many nurserymen and gardeners in England and Holland numerous interspecies hybrids. The hybrids had been precisely pollinated and the Dean's records were meticulously accurate. The earliest hybrids were of *G. cardinalis*, *G. tristis* and *G. carneus (blandus)*, the later ones included *G. natalensis* and *G. oppositiflorus* in their parentage. It is sadly recorded that many of the hybrids produced by William Herbert in the period 1806–16 were sterile, i.e. produced no seed, so that their propagation was limited. As the hybrids produced few cormlets and seldom split the corms to offshoots, these lovely hybrids died off. It is recorded that the fragrant ones — those with *G. tristis* parentage — were the first to succumb.

Figure 2 G. colvilleii 'The Bride'

We have confirmation of this tendency from the garden of Professor Barnard, who during the 1930–50 period produced in Dorset many winter species hybrids of *G. tristis*, *G. gracili*, *G. orchidiflorus*, *G. carinatus* etc. These beautiful little flowers known as the 'Purbeck Hybrids' gained Awards of Merit from the Royal Horticultural Society. It was found totally non-commercial to keep them going because of difficulty in propagation.

There is however a success story of a *G. tristis* × *G. cardinalis* hybrid. In 1823 a nurseryman of King's Road, Chelsea, in London, produced the first such hybrid that was fertile, and in the fashion of the times called it *Gladiolus colvilleii*, the hybridist being James Colville. The original *colvilleii* were red or pink, reflecting the *G. cardinalis* parentage. In 1926 James Colville put on sale *G. colvilleii alba*, a white sport with pale-pink anthers. In 1871 another nursery in London introduced *G. colvilleii* 'The Bride', an all-white sport. Corms of this hybrid are still available today under the same name — a tribute to the health, hardiness and reproductive powers of this hybrid.

The next series of breaks-through occurred about 1880–92 in both Holland and Britain. A successful crossing of *G. cardinalis* with *G. blandus* by the nursery of Schneevoogt in Holland gave a range of dwarf hybrids in colours from light pink through to lilac rose. These hybrids were given the name of *G. ramosus* and advertised as 'species', but later they became known as 'Charm' hybrids and are available as such today.

Krelage, also in Holland, back-crossed *G. ramosus* with *G. cardinalis* to produce a half-hardy, early-flowering dwarf cultivar. It is from this series of crosses that the *G. nanus* cultivars such as 'Spitfire', 'Nymph' and 'Blushing Bride' were produced.

There is some uncertainty as to the next break-through, particularly in regard to parentage. This was the so-called *gandavensis* hybrids (a Latinised form of 'from Ghent' in Belgium). These hybrids were produced by H.S. Bedinghaus, who left his employment as head gardener to a Belgian duke to set up his own nursery near Ghent. Although the originator claimed that *G. cardinalis* was a parent, the colour and floret pattern do not support this. Most botanists believe that the summer-flowering *G. natalensis* or *G. oppositiflorus* were the parents. Unlike the *nanus* hybrid group, the *gandavensis* hybrids were large flowered, tall, many budded and flowered in Belgium in September, if records can be relied upon. It is difficult to be sure of the parentage since at the time when Bedinghaus was doing his pollinating his *G. natalensis* was called *G. psittacinus* and his *G. oppositiflorus* was *G. floribundus*.

Close-neighbour gardeners in France and Holland took the *gandavensis* hybrids and developed them extensively to become the forerunners of our modern summer gladioli. In 1853 when Queen Victoria and Consort Albert visited Paris, they saw some of these development cultivars. Queen Victoria asked for some to be sent to the Osborne House gardener to be grown there. Ever willing to be seen as co-operative with British royals, the French sent a number of varieties with suitably royal names (changed from the original French) and from Osborne House they went to Kew and the Chelsea Physic Garden. To this day the French claim it was they who started off the gladiolus in England.

The validity of this claim is strongly disputed by the author since research has shown that before 1853 a stout-hearted Britisher called William Hooker had made a viable cross in England using what was then *G. blandus* and *G.psittacinus hookerii* —today *G.carneus* × *G. natalensis.* John Standish, a nurseryman in Bagshot, Surrey, also produced at about that time crosses similar to the *gandavensis.*

Standish had supplied the Belgium corms to Osborne House, and rumour has it that he supplied some of his own hybrids at the same time. This is believed to have come to light subsequently when some flowers of a deep-red and rather trumpet-like ('amaryllis–hippeastrum-like' was the phrase) were noted in the garden. It transpired that Standish has used pollen from *G. cruentis* on the Belgian stock. The British version of the *gandavensis* ultimately became known as the *brenchleyensis* (after Brenchley, the home town of William Hooker).

The French came back strongly in about 1877–82 with some interesting developments by Victor Lemoine of Nancy. Inevitably these became 'Nanciensis' hybrids and were quite distinct from the original *gandavensis*. The major difference was that the colour range was extended and many cultivars had clear white throats, replacing the darts of *G. cardinalis*. During his hybridising period Lemoine had introduced more than 100 distinct variations in his Lemoineii-Nanciensis hybrids. He had obtained lilac, bluish purple and lavender petal colour, deep-red, blood-red and carmine throat marks, and finally maroon, dark brown and deep green. This had been done by pollinating with *G. purpurea-auratus* (now *G. papilio*), *G. cruentis* (deep red) and *G. dracocephalus* (green and brown).

The *G. dracocephalus* hybrids gave the first smokey 'brown and green'-flowered hybrids great importance and significance. The best of the 'draco's' was named 'Pelletier d'Oisy'. Descendents of 'Pelletier d'Oisy' were used in Holland as recently as 1946 to create brown and green primulinus cultivars in their postwar revival of the gladiolus industry. It was from Lemoine's origination that the deep-purple, lilac and lavender cultivars arose in Germany.

In Germany it was Max Leichtlin who took up an interest in the work of Lemoine in about 1877. One of the Lemoine creations that particularly excited Leichtlin was the first deep true blue-violet cultivar, later named 'Baron J. Hulot'. Leichtlin used this and hybrids of similar blue, violet and deep purple shades and started a new line of hybrids — and naturally these became the 'Leichtlinii hybrids'. He also crossed the original *gandavensis* hybrids with the deep-red, trumpet-flowered *G. cruentis*. This later line of *cruentis* hybrids were never exploited commercially by Leitchtlin but via a French commercial organisation. He sold them in their undeveloped state and the stock eventually arrived in the possession of a Mr Lewis Childs of New York. Childs progressed the development of this line which he introduced as 'Childsii hybrids'.

The habit of this period of developing inter-species hybrids and giving them species status under the originator's name, i.e.*G. mellierii, G. leichtlinii, G. childii*, infuriated the professional botanists and confused the entire gladiolus nomenclature and classification. The procedure of giving location nomenclature i.e. *G. gandavensis, G. brenchleyensis* etc. was equally unhelpful as it also gave no indication as to the plant characteristics or colour. Nevertheless, the inter-species hybrids referred to were widely distributed in the United States under these names until about 1895 when American hybridists took over and extended the development.

Dr W.W. van Fleet, an American of European extraction, took a

Child's cultivar and back-crossed it with *G. cruentis* giving what he called *G. princeps*. Several other American hybridists engaged themselves in increasing the range of colour and flower size of the former European-created hybrids. It is true to say therefore that the bulk of the modern gladioli raised in the USA have as their ancestry the South African wild species in a handsome mixture cocktailed in Europe.

One American whose hybridisation is of great significance in the development of the modern gladiolus was the late A.E. Kunderd, who was born 1866 and died in 1965. His work produced the first ruffled florets in 1903 and the orchid-flowered, laciniated and double-floreted forms by 1923. This was a result of selective breeding rather than the introduction of new species pollen. During part of the lifetime of Amos Kunderd, Phillip O. Buch, also American, was raising other inter-species hybrids using more recently discovered species from the Cape (of South Africa) like *G. orchidiflorus*, *G. alatus* and *G. caryophyllaceus*.

Meanwhile, back in Europe, the Lemoineii hybrids and some of the more desirable of the hybrids of Dean Herbert's originations were being cleverly mixed in Somerset with some of the French hybrids of M. Souchet. This quite extensive pollination and seeding programme was the creation of Mr James Kelway of Langport in Somerset, where the family nursery of Kelway's was based. The first generation of hybrids were called *G. kelwayii*. The *G. kelwayii* of this first generation were rather 'the mixture as before', a fairly large-flowered wide-open type of floret some 4 to 5in (10–12cm) across the widest part, and placed widely spaced on a long stem. The colour range did not include white, yellow or orange.

On a botanical expedition around 1890 a summer-flowering gladiolus species had been discovered and sent to Kew. This species was sent for trial in that year at Kew and by 1902 was being widely distributed and sold under the name of *G. primulinus* (indicating primrose colouring). Capt. Collingwood-Ingram is quoted as saying that the *G.primulinus* was in fact *G. nebulicola* from the Victoria Falls area. The salient features of this species were the hooded floret and slender, pliable stem. The colour was either yellow or yellow with orange veining and 'clouding' — hence *nebulicola*.

Whatever it was and whatever its true identity, it was this species that James Kelway introduced into his breeding programme. The subsequent pollination of *G. kelwayii* with *G. whatever* gave rise to a new race of hooded and triangular-shaped flowers on tall, slender stems. These new hybrids were commercialised by Kelway's under the name 'Langprim hybrids'. The new strain had in many ways improved on the French hybrids by expanding the colour range to include light red, orange, salmon yellow, cream and ivory. In most of the flowers there were throat marks lighter than the main colour and darts of a different colour on the three lower petals. This is believed to be the first race of primulinus hybrids.

Mr Frank Unwin of Histon in Cambridgeshire was almost concurrently with Kelway working on the primulinus hybrids using *G. primulinus/G. nebulicola* as both pollen and seed parent. Kelway has used mainly the species pollen. Thus, Frank Unwin had created 'prims' of a much more delicate style, smaller flowered, consistently hooded, early flowering and, above all, being less complicated genetically propagated well and remained healthy.

It is now a fact of history, that 'Langprims' went out of favour and the 'classic prim' to the Unwin specification became the accepted 'primulinus hybrid' as recognised for exhibition purposes by both the Royal Horticultural Society and the British Gladiolus Society when it was formed in 1926. Unhappily, although the primulinus hybrid form of the summer-flowering gladiolus was to become very popular in Britain, its acceptance in the USA, Canada, Australia and New Zealand was less than enthusiastic. Presently, only in Britain does the National Gladiolus Show include classes for the primulinus hybrids.

In Holland, using parent gladioli from England, some Dutch hybridists revived the primulinus after World War II, notably Messrs De Bruyn of Noordwijk and Piet Visser of St Pancras. Some amateur British hybridists continue to raise new and improved primulinus hybrids for exhibition purposes, notably E.W. Brown of Formby, Lancs.

In Germany at this time the family firm of Wilhelm Pfitzer at Stuttgart were intensively engaged in improving the summer-flowering gladiolus. Using existing commercially available cultivars they set out to produce a robust and healthy strain of large-flowered cultivars in a wide colour range. From their extensive inventory they selected about ten cultivars for commercialisation and export. Amongst this ten were 'Snow Princess' 'Commander Kohl', 'J.S. Bach' and 'Rosemary Pfitzer'. The cultivars achieved worldwide acclaim for health and propagating power, and were the mainstay of the floristry trade in their day. Additionally, these cultivars were used extensively in breeding in Holland, Britain, Canada and USA. As an indication of the durability of the Pfitzer cultivars, 'Snow Princess' is still being grown commercially 60 years after its introduction. Its nearest rival is Carl Fischer's 'Friendship' which is still available after 48 years.

By the year 1910 there were available literally thousands of cultivars of the summer gladioli from innumerable sources. The colour range, size of floret, number of buds on a stem and season of flowering had all been expanded by selective breeding and intensive propagation. From 1910 onwards very little work involving the introduction of new species was being done anywhere in the gladiolus world. Instead, the attention of most hybridists was turned to the production of bigger and better specimens of the new 'grandiflora' gladiolus cultivars. Others however had ideas of producing new types of gladioli from what amounted to a remix and reselection of the existing cultivars. Selective crossing and the occasional mutation gave rise to some rather attractive new types of gladiolus in the smaller-flowered groups.

The 'primulinus hybrids' continued to be developed in Europe but seemed not to be fully accepted in North America, though the small-flowered or miniature gladioli were. The large-flowered gladioli were being grown in quite large numbers in the 1920–30 era, and in many towns in Britain flower shows began to include specific classes for gladioli where previously the few gladioli seen at the

*Figure 3* Gladiolus natalensis

shows were exhibited as 'summer flowers' or 'bulbous perennials'. Messrs Kelway in Somerset and the Mair family from Prestwick in Scotland were hybridising and distributing some very progressive cultivars, and Unwins of Cambridge had become established as the primulinus suppliers. In 1926 the specialist British Gladiolus Society was formed and in 1927 held its first specialist show devoted exclusively to gladiolus hybrids and species.

The years from 1930 to 1939 saw very little progress in the production of new types of gladioli though quite rapid progress was made in the improvement of size and form. Professor Palmer in Canada produced some selectively line-bred cultivars which became the forerunners of the present show gladioli. A very significant milestone on the road was 'Picardy', which was proclaimed the 'Gladiolus of the Century', its pollen giving rise to some wonderful gladioli worldwide.

For the European gladiolus enthusiast the years 1940–47 were very lean ones. Gladiolus and other flowers took a back seat in the horticultural wagon whilst all efforts went to food production during World War II and the immediate aftermath. Fortunately for the Europeans some good gladioli were still to be found in USA, Canada, Australia and New Zealand, whilst South Africa still had extensive cultivation of pre-war Dutch cultivars and the native species.

The Dutch firms, Konynenburg and Mark, Salman, Visser, De Bruyn and Preyde, to name but a few, searched the Low Countries and France's rural areas for vestigial plantings of the Dutch cultivars that could have survived five years of neglect. Friends in North America and Australasia sent seed of gladioli, and the authorities co-operated magnificently by allowing cormlets and small corms to be imported to recreate the gladiolus industry in Europe. The South African government's assistance in permitting their variants of *G. psittacinus* (as they were then) to be sent to Holland was welcome and effective.

The so-called *nanus* hybrids had survived in Europe in greater numbers than the *grandiflorus* so that the earlier revival of hybrids from these (by crossing with *G. primulinus* and *grandiflorus* hybrids) created some attractive new, small-flowered gladioli. About 1948 Unwins of Cambridge exhibited the first Peacock hybrids — *nanus* × *primulinus* hybrids. These were halfhardy and early flowering, as would be expected from the parentage.

By using some of the small-flowered cultivars from America and Canada as seed and pollen parents in crosses with their own small-flowered cultivars, hybridists from Konynenburg and Mark in Holland produced a series of 'Butterfly' hybrids. It is frequently stated that the *G. papilio* species were used in the Butterfly cross, but this is denied by Arie Hoek, who did the pollination. The Butter-fly gladioli were very well received at the Gladiolus Competition organised by the Royal Horticultural Society in London's Westminster, and the cultivar 'Ares' was awarded both Certificate of Commendation and Award of Merit. Later 'Green Woodpecker', perhaps the most popular Butterfly, was similarly honoured.

A chance mutation or sporting in some hybrids derived from 'Brightsides' × 'Roi Albert' gave the hybridist Mr Leonard Butt of

Ontario, Canada, the break-through of a lifetime. Though both 'Brightsides' and 'Roi Albert' have plain petals, the sport from the crosses produced small-flowered gladioli with intensely ruffled or crinkled edges to the petals. Three cultivars were selected and stabilised. These were 'Bo-Peep', 'Crinklette' and 'Emily's Birthday'. Subsequently, seedlings from crosses made between the first three cultivars also gave a high proportion of ruffled or crinkled-petalled plants. So were born the 'Ruffled Miniatures' so popular with the florists and flower arrangers.

The years 1950–60 were possibly the decade of the hybridists who were concentrating on improving existing types of gladioli hybrids. A classification system and a points system for judging exhibition gladioli had been established for *grandiflorus* gladioli of all sizes. The ideal show gladiolus had been specified in every detail, the hybridist now had a blueprint for the perfect 'glad'. Biologists were also deeply involved in perfecting cultivation techniques and chemical agents to keep the 'perfect gladioli' healthy and productive. New propagation techniques like tissue culture (meristem) had been successfully applied to lilies and orchids, and the gladiolus was to be the next candidate

*Figure 4* G. callianthus

In America hybridists like Larus, Baerman, Fischer, Roberts and Frazee were introducing each year some magnificent exhibition large, medium and miniature-flowered cultivars, some of which are still winning prizes 25 or more years after their introduction. In Canada, it was Butt, Rich, White, Jack, and Palmer. A new generation of hybridists are doing today for the gladioli what these pioneer hybridists did in their time, by improving each year, minutely but perceptibly, on the great gladioli of the 1950s and 60s.

In Holland there are hybridists like Snoek who has introduced about 20 superb new show cultivars in the last five years. With Preyde, Visser and the Konynenburg and Mark team, some hugely successful show gladioli are coming from Holland. Britain has three hybridists whose creations have achieved international acclaim. These are John Pilbeam of Sussex — originator of 'Red Smoke' and 'Inca Queen' — and Eric Anderton of Lancashire with the cultivars 'Holcombe', 'Alex Back' and 'Alex's Sister.' These two hybridists produced large-flowered cultivars, and the third British hybridiser is Eric Brown of Formby, Lancashire, who specialises in primulinus hybrids. His cultivar 'Rutherford' is a classic exhibition 'prim' which won the coveted 'Gold Vase' Trophy at a British Gladiolus Society Exhibition. It is a very attractive if unusual colour: rust red-veined silver.

It would require a library of catalogues to detail the contemporary history of the gladiolus since 1960. It is really the continuing story of dedicated hybridists striving for the perfect gladiolus — as ever. The most significant step forward and highlight in gladiolus history of recent times was the work of Mrs Joan Wright in New Zealand in 1955. Along with many other hybridists Mrs Wright was trying to create a fragrant gladiolus by crossing a garden hybrid with a fragrant species. It is generally accepted that the species with the strongest fragrance and the summer-flowering characteristics is *Acidanthera bicolor*. Mrs Wright made something approaching 200 crosses on small-flowered summer hybrids using the pollen of *A. bicolor*. Over 220 plants eventually flowered and of these only one flower had fragrance. This was a cross with the cultivar 'Filigree', and proved to be consistently fragrant for up to 48 hours per floret. It was distinctly fragrant with the characteristic fragrance of the *Acidanthera*; however, the floret characteristics and the plant style was intermediate between *Gladiolus* and *Acidanthera*.

This $F_1$ cross was again pollinated with *Acidanthera* and the resultant hybrid was fragrant and compatible with most summer-flowering hybrids. In 1957 the hybrid 'Lucky Star' was trial tested in Europe and found to be tetraploid (four times the normal chromosomes, i.e. 60) and compatible in most cases with summer gladioli. 'Lucky Star' was considered at the time to be bi-generic, that is a hybrid of two distinct genera, and was therefore called a *Gladanthera*. Subsequently, following extensive work by botanists in South Africa, and the publication of *Gladiolus — A Revision of the South African species* by Dr Lewis, the genus *Acidanthera* has now been re-classified as *Gladiolus callianthus*. Mrs Joan Wright continued her work with 'Lucky Star' in a hybridising programme with the objective of retaining the fragrance of 'Lucky Star' but modifying the colour and shape of the floret. After a few years of very limited progress Mrs Wright passed on her stock of Gladanthera and devoted her time to breeding horses.

The author was fortunate in obtaining stock of 'Lucky Star' as also were Dr A.P. Hamilton of Sussex and Ing.Igor Adamovic in Bratislava, Czechoslovakia. The first exercise undertaken with 'Lucky Star' was to confirm the chromosome number as n=60, that is tetraploid, which it is. Secondly, to select a number of coloured miniature-flowered culitvars of gladioli which also had chromosome number n=60. Although it is generally believed that most garden gladioli are tetraploid, this is far from true as a great many are triploid (n=45).

It was strongly felt by one or two botanists that the key to the coloured fragrant *G. callianthis* × garden hybrid was to establish a tetraploid *G. callianthus*. Currently available *G. callianthus* (*Acidanthera* in 1970) at that time were from three sources: the wild Ethiopian *A. bicolor*, *A. bicolor murialae* (Kelway) and *A. tubergenii* 'Zwanenburg' (all three of which are now *G. callianthus*). A series of experiments and tests were carried out in the years 1972–6 to find a consistently tetraploid version of *G. callianthus*.

Chromosome numbers can be changed, it is stated, by gamma-ray bombardment, irradiation by isotopes and by treating growth points with colchicine. All three ways were tried, and finally by a method we do not wish to disclose carried out by a person we do not wish to name, there is now a stock of 40 or so flowering-size corms of a confirmed tetraploid *G. callianthus*. The strain of *G. callianthus* is compatible with 'Lucky Star', and tetraploid *G. callianthus* is being used extensively in a programme in Czechoslovakia by Igor Admovic in Bratislava. His greatest success to date is 'Eau Sauvage' a light-pink and significantly fragrant Callianthea, the seed parent of which is a garden hybrid.

In England, apart from the tetraploid *G. callianthus*, we have tetra-*callianthus* × red *primulinus*, several *callianthus* crosses with cream and yellow miniature cultivars and some pale-rose slightly fragrant hybrids from *G. callianthus* × cv 'Gigi'. The problem that exists is to breed out the tubular floret shape of the *callianthus* and dilute or eradicate the rather unattractive maroon or purple throat patch.

Alongside the search for fragrance via *G. callianthus*, much work has been done in America in the past 20 years by hybridists not using the species but relying exclusively on the sporting of fragrance in normal garden hybrids. The work of Spencer and the Rev. Clifford Buell is of historical significance in this regard, and the rose-like fragrance of 'Acacia', 'Yellow Rose' and 'Cliffie' produced by hybrids of these gentlemen's creation may well be the 'added ingredient' that will bring us the real fragrant gladioli that future historians will record.

## North American Developments

The development of the summer-flowering gladiolus in North America (i.e. USA and Canada) over the past 80 or so years has followed a rather different pattern from that in Europe. Our trans-atlantic cousins have tended to concentrate on producing gladioli in quantity and of quality such as to give an economic return in financial terms. The early American gladioli growers and hybridisers like Mr Childs and Mr Kunderd not only developed gladioli, they also commercialised them. Mr Childs was publishing a list and offering his *childsii* hybrids (improved *gandavensis* hybrids) in 1894 when his newest introductions were selling for up to $5 per corm — real money in those days. A.E. Kunderd began to develop his own strain of garden gladioli hybrids as an amateur in 1880 when he was quite a young boy. In the following years the Kunderd Farms at Goshen, Indiana, grew until they covered 27 acres (11 ha) alone of gladioli of Kunderd's raising, representing over 5,000 different varieties.

Some of the Kunderd historical highlights will indicate the importance of his contribution to the gladiolus industry in the USA:

1919: 'Indian Summer' — a cross between *G. kunderdii* and a species *G. quartinianus* (now a *natalensis* group species) was introduced at $25 a large corm. This cultivar had a 17-bud stem with 5 open ruffled florets in a lilac-pink shade.

1922: 'Kunderd's Glory' — the first ruffled gladiolus that gave up to 80% ruffled seedlings. Price $10 a corm any size.

1923: 'Kunderd's Lacinatus' — the first commercial cultivar with the laciniated petals. Offered at $1,000 a corm and he actually sold five to one client.

1924: Kunderd's sensational 20-bud spike. Won a trophy outright at Kalamazoo and over $2,000 in prizes. Sold in 1924 at $100 a corm. Plain-petalled medium-sized pink.

1927: By this time Kunderd was listing 30 primulinus hybrids, 27 plain-petalled and 30 ruffled cultivars at an average price of $3.50. A ruffled black-red was named by permission of Thomas Edison and cost $100 each.

A.E. Kunderd spent over 80 years raising and growing gladioli until his death in 1965.

Another American, Phillip O. Buch, pioneered the species hybrids in USA, developing his own strain of Buch's *alternatus* hybrids at his home in New Jersey in the period 1935–52. The Buch hybrids were similar to *G. tristis concolor* in form and size but in a wider colour range. The florets had three outer petals long, large and pointed, and the three inner petals were small and rounded. The overall effect was of an alternate petal grouping, and where the inner petals differed in colour from the other three the contrast was dramatic.

From the 1950s onward several hybridists took up the challenge to develop new commercial and exhibition gladioli. The most notable of these hybridists were Baermann, Fischer, Roberts, Frazee, Turk, Larus, Greisbach and Walker. These gentlemen between them have produced over 75 per cent of the registered American gladiolus.

Canada's contribution is quite significant, starting with Dr Palmer, whose cultivar 'Picardy' was the sensation of its time. 'Picardy', with other similar line-bred seedlings from the same parents, formed the basis of the revival of the large-flowered gladiolus in Europe in the late 1940s and was extensively used worldwide by hybridists. Leonard Butt of Ontario created, introduced and popularised the ruffled miniature with his seedlings 'Bo-Peep', 'Statuette' and 'Crinklette'. Some magnificent exhibition cultivars originated in Canada in the years 1955–75 from hybridists such as Jack and White, who will always be famous for 'Violet Charm', 'Salmon Queen', 'Landmark', those giants of the show bench. Alex McKenzie of Ontario is currently producing some superb exhibition material to carry on the Canadian tradition. Cultivars like 'Drama', 'Incomparable', 'Summer Special' and 'Diva' will take their place in the Courts of Honour of the future.

## William Herbert — the Great Pioneer

The Very Reverend and Honourable William Herbert LLD, BD was born on 12 January 1778, being the third son of the First Earl of Caernarvon, by Lady Alicia Maria Wyndham, eldest daughter of Charles, Earl of Egremont in Cumberland. This remarkable man was one of the most learned and talented men of his time, and though his accomplishments were of the highest order he remained a modest and generous person throughout his life.

He was educated at Eton, graduated at Oxford and later won a doctorate. He married in 1806, his bride being the young and beautiful second daughter of Viscount Allen, the Hon. Letitia Allen.

*Figure 5 The Very Rev. and Hon. William Herbert LLD, BD. Reproduced from
an engraving published in the* Gentleman's Magazine *in 1847*

A devoted family man, Herbert was prominent and competent in
many academic fields. He was also a classical and theological
scholar, expert in Greek and Latin and fluent in modern European
languages. Additionally he was honoured for his many scientific
achievements. In politics he sat as Member of Parliament for
Hampshire in 1806 and for Cricklade in 1811. In 1814 he was
appointed Rector of Spofforth in Yorkshire, and his ecclesiastical
career extended to his appointment as Warden of Manchester and
later Dean of Manchester. It was during his term of office as Rector
at Spofforth and later at Mitcham in Surrey that William Herbert
grew his gladiolus species and hybrids.

He was an eminent botanist, famous for his knowledge of
bulbous plants which he grew in quite large numbers. Amongst his
many literary publications were several items in the *Botanical
Register* and a long series in the *Botanical Magazine*. He addressed
the Royal Horticultural Society on several occasions and contributed
to that Society's publications. His standard volume on *Armaryl-
lidacae* was issued in 1837 and his 'Crocorum Synopsis' graced the
*Botanical Register* in the years 1843-4-5.

The year 1842 was perhaps the peak of Dean Herbert's achievements not only in horticultural matters but in natural history generally, for in that year he was honoured by the Royal Horticultural
Society and was also elected President of the British Association.

Dean Herbert was a prolific writer on many subjects, though
perhaps his botanical treatises were the most famous. His extremely
valuable contributions on hybridisation of many species were highly
respected, for they were based on personal experience and the
meticulous records he always kept. His work on *Dianthus* species
and hybrids and the inter-species crosses of *Gladiolus* are classic
examples. The *History of the Species Crocus*, originally published as
articles in the *Journal of the Horticultural Society* in London, was
reprinted shortly after his death and dedicated to his memory.

Dean Herbert's work on the species of *Gladiolus* began almost coincidental with his marriage and at a time when his political career was launched. He started by increasing the stock of species he had acquired from South Africa and later by using forms of the species to produce his famous inter-species hybrids. The species were *G. tristis*, *G. carneus*, *G. cardinalis* and *G. flordibundus*.

During the period 1806–9 at least 32 different inter-species hybrids were created and recorded by Dean Herbert. After 1809 further crosses were made using *G. caryophyllaceus* and *G. liliaceus* on the *tristis* hybrids. At this time he had been given or had acquired several variants and sub-species of *tristis* with different flower colour, and this programme of breeding had created great interest and enthusiasm for fragrant hybrids. By 1837 William Herbert had several dozen large groups of *Gladiolus* species and hybrids growing in his garden and additionally was growing his fragrant *liliaceus* and *caryophyllaceus* hybrids in a 'stove house'. (This was the current terminology used to describe a large, tall, conservatory-type glass house heated by a coal or wood-fired boiler circulating either ducted hot air or hot water through radiator pipes).

The first really fragrant hybrid was recorded by both Colville and Herbert in 1823 though there is no information as to whether the two hybrids had the same parentage. Seedlings from these crosses were sold from Colville's nursery in Chelsea in 1835. One item was auctioned — a rose pink fragrant *cardinalis* × *tristis* — and was bid up to 'several guineas'. In more recent times the catalogue of Colville's sale also went for the modern equivalent of several guineas.

As if all his achievements were not enough, William Herbert enjoyed the reputation of being a 'good shot', always to be relied on 'for a couple of brace of birds on any shoot'. Much of the Herbert *Gladiolus* material was disbursed in the years 1842–6 when through age and ill-health the great man could no longer continue his work. Dean Herbert died at his mansion in Herford St, Park Lane, London, on 28 May 1847 at the age of 69.

William Herbert made a unique contribution to the development of the modern gladiolus cultivar. In many ways it is unfortunate that his work was not more widely recorded and preserved. His memorial must surely be the thousands of hours of pleasure that his work inspired in the lives of the present-day gladiolus enthusiasts.

*Acknowledgement*: Miss Angela Evans for her meticulous research into the non-botanical aspects of 'the Great Dean'.

*Figure 6*
G. caryophyllaceus: *carnation-scented species used to produce fragrant hybrids.*

# *Chapter 3*  **Propagation**

We all can purchase mature corms from a specialist supplier who in turn obtains them wholesale from a commercial grower, usually from America or Holland, but eventually the keen enthusiast will require to propagate his own for various reasons. Perhaps having purchased basic stock of a new and successful cultivar, which may incidentally have cost quite a bit of hard-earned cash, it would certainly be wise and economical to produce your own corms instead of resorting to a bank loan to purchase more. It certainly is expensive to purchase totally new requirements each season.

Once a cultivar has proved its worth in your own particular environment and shown to perform well and be free of disease and obvious virus, from a good 'strain', as gardeners say, then that stock should definitely be retained. Another good reason is that existing stocks are already acclimatised, and corms propagated from these stocks and grown again in the same conditions certainly have an advantage over corms that have been grown in far different situations, in far warmer climates and in faraway places. We can also practise what is termed 'selection' and 'reselection' of stocks, but more about this later.

I like to plan my future programme well in advance, and by growing my own I can then propagate the cultivars that I require and in the correct numbers, and also have them available at the right time for planting. This is important in the north of England and in Scotland, as corms purchased from America sometimes do not arrive until mid or late April. It is very frustrating to be waiting to see if the order has been fulfilled and being dependent on transportation problems and substitution. Usually it is just that particular cultivar that you have looked forward to growing that year that is not in the parcel, and has been replaced by something you did not want in the first place. Then there is a certain satisfaction from producing your very own and harvesting a good crop of corms, only equated and perhaps surpassed by those wonderful blooms in the summer. The very fact that gardeners are gardeners is that we like to be part of the natural process of production and reproduction. Above all, if you breed gladioli, then you must certainly be able to propagate your own corms from the selected seedlings. It is no use breeding that potential world beater, or the one that is the fore-runner of a completely new line as yet unavailable, if it cannot be reproduced!

This chapter is not devoted to the methods of the commercial growers with their vast acreages and huge outputs, but to the hobbyist and enthusiastic amateur. That said, many of the techniques have been adopted from the successful practice and research of the large professional growers, institutes and research stations throughout the world.

Now before I get letters from all my specialist supplier friends complaining that I am taking away their very livelihoods, let me say that they do a great job! Yes, they do enable us to grow all those new creations from all the talented hybridisers from around the world. Yes, and I do know that it is not easy for them to avoid substitution all the time, being as they are in the hands of and depending on the growers, and because of the problems of crop failures; and having to crystal-ball gaze to anticipate demand, exchange rates, changes in importation taxes, postage rates and the like.

## Basic Propagation

The way in which gladioli are normally propagated is by cormlets. These cormlets, which are really small immature and miniature corms, are formed at the base of the corm on short stolon-like shoots. They will be evident on lifting the corms in the autumn for storing during the winter (see Figure 6).

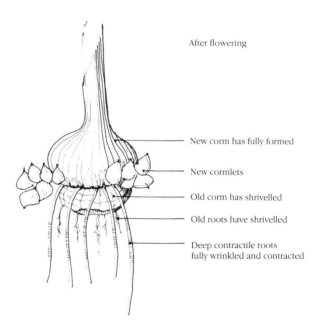

After flowering

New corm has fully formed

New cormlets

Old corm has shrivelled

Old roots have shrivelled

Deep contractile roots
fully wrinkled and contracted

*Figure 7  A corm with cormlets*

The cormlets do require growing on for a year or two during which time they will increase in size, and until they are mature corms capable of producing a full-size flowering spike. They way they are treated and grown will obviously to a great extent determine the quality of the corm and also the flowering spike that is produced. Harvesting, curing and storing must be carried out in the same manner as recommended for corms. They do keep well in storage however, being protected by a hard shell. The basic method of propagation, in which identical reproductions are made of the chosen cultivar, is really rather simple and straightforward, but I will describe in detail how the very best quality corms can be produced.

# The Size and Number of the Cormlets

The size and number of the cormlets found on each corm can vary tremendously and a few factors are responsible for the differences. Some cultivars naturally produce very few and other cultivars an amazing number, even in some cases totalling up to a hundred or so. The size can also vary from perhaps the size of a rice grain to a walnut. However, you can have some control and effect on the reproductive qualities, especially desirable when the cultivar is known to be 'shy' or it is necessary to increase the stocks of a desirable cultivar or seedlings as quickly as possible. Obviously, if the plant can be encouraged to flower early in the season, leaving time in the autumn for the corm and consequently the cormlets to mature, so much the better.

Most cormlets, and the biggest, are produced from stock plants, that is those that are being specially grown for the purpose of propagation. They are planted as early as possible and either not allowed to flower or the flower spike is removed as soon as colour shows. The reason that it may be desirable to first see the colour is to check that it is of the correct cultivar and not a 'stray' or 'rogue'. All the energies of the plant will then be directed in building up the corm and also the size and number of the cormlets.

It is a fact that corms planted very shallow also produce more and bigger cormlets, but there is a very special technique that can be employed to get the absolute in size and number of cormlets from a corm. Before planting, a corm can be cut into sections or slices which are each separately planted and will each produce its own quantity of cormlets. In fact, due to a rejuvenation process, this increase in total weight of the cormlets can be up to 25 times greater! Old mature corms are best for this purpose, having more shoots, buds or eyes on the corms, even some of them immature. The corms are then cut into slices, but it is important that each has a shoot, bud or eye at the top and a portion of root scar with root buds at the bottom. (See Figure 7).

It is advisable to sterilise the knife before the surgical operation of cutting each corm, to prevent the spread of any possible virus or other infection. The cut portions are then treated with a suitable fungicide powder and are then planted shallowly as early as possible in the spring. I plant them in large pots of soilless compost and in the protection of a greenhouse. You will be amazed on lifting in the autumn at the total quantity of cormlets produced by one corm. I have used this method many times to increase the stock of a new seedling that I have bred, and it also is a satisfactory way of utilising an old corm that is probably past its best as regards its flowering capabilities.

# Selection of Stocks for Propagation

It is absolutely essential that selection of stocks for propagation should receive utmost consideration. Only the best quality material should be used, that is, disease and virus free. Corms that have produced the best flower spikes should be marked and retained and all others immediately eliminated from propagation or 'mother'

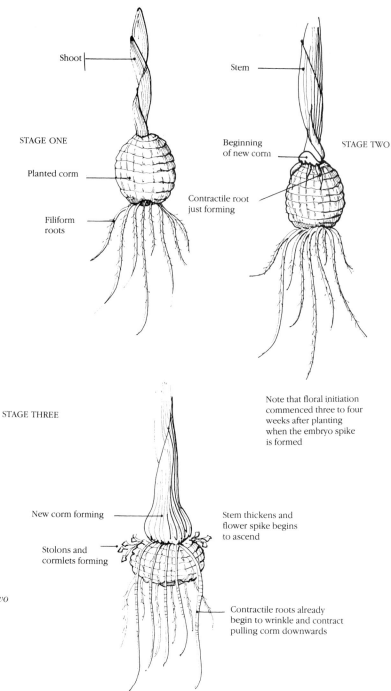

Shoot

STAGE ONE

Planted corm

Filiform
roots

Stem

Beginning
of new corm

STAGE TWO

Contractile root
just forming

STAGE THREE

Note that floral initiation
commenced three to four
weeks after planting
when the embryo spike
is formed

New corm forming

Stolons and
cormlets forming

Stem thickens and
flower spike begins
to ascend

*Figure 8 The three
stages of growth: Stage
One, approximately two
weeks after planting;
State Two,
approximately seven
weeks after planting;
Stage Three,
approximately eleven
weeks after planting*

Contractile roots already
begin to wrinkle and contract
pulling corm downwards

stocks. From a batch of say 100 corms of one cultivar, it is far better to select as few as one or two that have performed best than to retain a far greater number that are not so good. Continued selection and reselection is necessary and eventually the quality of the stock will be upgraded tremendously, and actually improved. This is where the keen and enthusiastic amateur can actually score over the commercial grower, who cannot really afford to practise too much selection but needs all the cormlets and stock he can obtain. It is just a matter of economics. Exhibitors of all other subjects such as chrysanthemums and dahlias, and certainly the giant onion and leek growers, all practise intense selection to gain advantage over their competitors and to develop their own 'strains'. This equally applies to gladiolus growers if success is desired.

At this point it may be useful to highlight a few points to enable us to determine which plants we should select out of a growing batch. Obviously the performance and quality of the flowering spike is very important. Long, straight spikes carrying the most bud count and with clear coloured florets with no streaking that could indicate virus are the ones to look for. All other things being equal, it is usually the plant that flowers first in a batch of the same cultivar that is best and most healthy. Leaves should be evenly coloured green with no streaks, blotches or changes of colour, and certainly not harbouring any moulds or rots. Comparisons should be made and it will soon become obvious which plants are the healthiest.

It is necessary to keep stocks insect free by regular spraying. Viruses can be spread by aphids as they can also be transferred by cutting knives unless precautions are taken. Blades should really be wiped and sterilised between cuts. Most commercial stocks of gladioli do carry virus although not always apparent to visual inspection. And it is really a matter of degree. Only by specialised laboratory techniques can virus be detected. But the observant amateur can again have the advantage over the commercial grower by being in the position to give detailed observation and attention to the comparatively few numbers. And here is the advantage that a breeder of gladioli has. Seeds do not normally transmit virus and diseases and therefore a clean start is virtually assured. Only those who have grown virus-free stocks can realise what tremendous growth they can make, with terrific flowerheads.

If you like to introduce new cultivars from others then try to purchase from a specialist who actually imports direct from the breeder or introducer of a particular cultivar. Get a few corms as soon as possible in the first year of introduction. Better to do this and propagate yourself from that basic stock, still practising selection and reselection, of course.

## Recommended Method

It must be realised straight away that the cultivation methods for the production of corms is far different from those used for the production of first-class blooms. In fact commercial growers, especially in Holland where they are subject to the most stringent control and inspection, are required to elect the purpose for which the crop is being grown, even before it is actually planted, to be granted the appropriate licences.

This section is devoted to describing a successful and proved method, suitable for the keen amateur enthusiast, for producing absolutely first-class stock. The high cost of purchasing good corms in the required and successful cultivars justifies quite an amount of preparation and care, the cost of the correct compost, fertiliser and other necessary items being, in comparison, minimal.

There are some of the more favoured parts of the British Isles where the climate is mild and the soil suitable to enable cormlets to be planted directly outside in drills just as one would sow a crop of culinary peas. The longer season of growth and warmer conditions would allow a reasonable crop of corms to be harvested, but even so only of certain cultivars. But in reality there are not many such locations. I don't think there are any commercial corm growers and producers in the British Isles, and the reason is obvious. If I am wrong then it would give me great pleasure to be proved so — but please forward to me the name and address immediately!

Living in the north of England, the Rossendale Valley of Lancashire to be precise, which is renowned for a very damp climate more suitable for the weaving and spinning of cloth, and the wearing of shawls and clogs, I can tell you that raising good gladioli corms is a challenge. To be quite honest it is not easy in this area either to propagate or even keep decent corms from one year to the next. Faced with this problem, I tried many ways of increasing stocks by growing on cormlets, and as I had become interested in breeding, by seed, I realised it was futile to breed gladioli if it was impossible to propagate them. I pondered the problem continually, and concluded that the climate and conditions in this country were just completely unsuitable compared with Holland and parts of America. But a factor in growing any type of plant really well is to study the natural habitat of the species from which the cultivated hybrids are derived.

Whilst the hybrid gladiolus as we know it today has a varied ancestral background, the general condition and lifestyle in the wild, especially South Africa, is something like this. The corm lies dormant in dusty, dry conditions during the hottest part of the year (summer), but when autumn arrives with the rain and cooler conditions the plant begins to grow and produce a flowering spike in the winter. The cycle is then completed when the plant dies down and matures to a dormant corm with the return to hot, dry weather.

But we have in effect now turned the seasons upside down. We plant the corms in spring and a flower spike is produced in the late summer or autumn, but after that time the weather is getting colder and wetter, especially in the north. In a poor autumn the corm never really matures. I then considered further aspects of the problem in relation to soil conditions and moisture and especially the bulb-growing districts of Holland. The soil is well drained and derived mainly from sand, and the water table is usually below the level of the corm. In other words, the ideal situation seemed to be a dry corm but moist roots. This would avoid any rotting of the corm and help it to mature.

Can I tell you a story? I was about to purchase my first and very own greenhouse having seen an advert in the local newspaper. I took my dad along, as he thought for advice, but the fact was he was loaning me the money. The retired army officer owner took us down the garden to view the prospective purchase. He had not grown anything in it for about five years and the greenhouse was empty, except that at the far end was a huge clump of gladioli in full bloom which I instantly recognised as the variety 'Picardy'. Yes, they were 'varieties' in those days and not cultivars! I was amazed by the beautiful, clean and vigorous growth of plants in all stages. Some were obviously first-year cormlets and others varied right through the stages to mature plants. The beds were raised by brickwork above the level of the ground but the soil was absolutely dust dry.

You bet I examined those plants closely when dismantling the greenhouse. Incidentally, I became the proud owner for £14. The gladioli had not been watered for five years; the corms were perfectly dry but the roots had travelled down to receive some moisture. Certainly a lesson learned and, of course, I repeated the experiment in my new greenhouse the following year. End of story but the beginning of many years of the development of my ideas.

My present system is something like this. It may seem to be a lot of trouble but well worthwhile. You will need a greenhouse, and ring culture is the name of the game. I make my rings out of plastic damp-proof course from the builder's merchants, but you could use old buckets with no bottoms and other inventive ideas. These rings are placed on the cultivated greenhouse bed and pot thick. Fill with a soilless compost or make your own with Chempak, peat and sand or perlite. Cormlets after peeling are planted say 10 to a ring about 2 in (5 cm) deep in the compost. The end of February is a good time to start in a cold house to ensure a long season of growth. Now water both compost in the rings and the soil below and contrive to do this until about July. Liquid feed can be supplied in the watering, then cease watering and feeding the rings altogether and only keep-ing the soil below moist, not wet. Admit plenty of ventilation and no shading, please. If any flower spikes do appear you can break them out. Any flowers produced would be poor; remember we are grow-ing for corm production and not flowers.

At the end of August, cease watering altogether and let both the rings and soil of the greenhouse bed become dust dry. The plants will still be green, but do not on any account apply water; please resist the temptation. Now here comes the exciting time. Sometime during October, hopefully if you can wait that long, break open the rings to reveal those super corms. They should be the best you have ever seen if all has gone well; clean, ripe, plump, dry and mature. Most will be ready for planting in your exhibition patch the follow-ing year. Some may require a further year depending on the cultivar.

Do remember the basics of insecticide and fungicide spraying and that glads do not enjoy growing in the same soil or compost each year. Use fresh compost and rotate the plantings on new soil in the greenhouse each year — the same rules as for tomatoes with regard to the soil sickness problems. Seeds and seedling corms can be grown on in exactly the same way. Eventually you should be able to establish a programme and system to enable nearly all your corm requirements to be produced by yourself! Think of the advantages.

## What is a Good Corm?

The biggest corm is not always the best, and remember vigour is usually on the side of the young. A young, plump corm, the depth being equal to the diameter and often termed a high-crowned' corm, is generally better than an old, flat corm, probably wise with age but in reality past its best (see Figure 8). The age of the corm can also be determined by examination of the base or root scar, something like the age of a tree by examination of the annual rings in the trunk. A young corm has a small root scar but an old corm has a large one. With practice a very accurate assessment can be obtained of the actual age or number of years from the cormlet stage.

However, it must be said that the performance of different cultivars does vary to some extent in relation to the size and age of the corm. It is a matter of being familiar with and having the experience of growing those cultivars in order to gain the maximum potential from them. There are certain ones that perform really well from very young corms but very poorly afterwards; there are some that perform really well from young corms and equally as well from older corms, within reason. Obviously a good corm is clean looking and has no disease marks or any other disfigurations. If the corms are produced as I have described and in a soilless compost mix of peat and perlite, then they should be perfectly clean. Good corms can also be grown in large-diameter pots, and I definitely prefer clay pots to plastic as to me the advantages are enormous. Control of watering is very important if pots are used, bearing in mind all my previous remarks on that subject.

*Figure 9 Examples of a good and a poor corm. The good corm, on the left, is young but mature and plump, with a high crown, one apical shoot and a small root scar. The poor corm, on the right, is old and too mature, being flat with many shoots and a large root scar.*

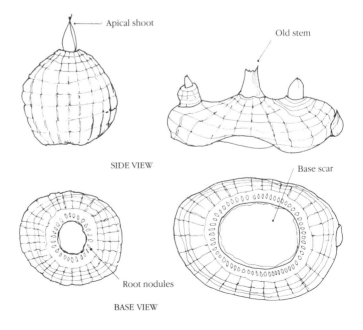

Apical shoot

Old stem

SIDE VIEW

Base scar

Root nodules

BASE VIEW

# Commercial Propagation from Cormlets

Now just a few words regarding the commercial propagation of gladioli from cormlets. As can be imagined, the number of corms produced is enormous and is directly related to the economics of the industry. Practically the whole of the operation and processes is mechanised, from preparation of the ground, planting, cultivation, spraying for disease and insect control, weed control, to lifting, drying, cleaning and ripening. The storage conditions are accurately controlled as regards temperature, humidity and air circulation. Fumigation, chemical dipping and 'hot water treatment' of stocks is also practised. Most of the commercial growers produce the culti-vars either for the vast cut-flower market or for general sale to gardeners who are not necessarily enthusiasts, or grow for exhi-bitions. There are just a very few, a minority, who themselves are enthusiasts and grow the unusual or exhibition cultivars.

# Commercial-scale propagation

Commercial-scale propagation of gladiolus cultivars is limited to those countries where very large areas of arable land are readily available. To be effective and economic the operation must be large also, running into millions of corms over several hectares of land; anything from 40 to 250 ha (100 to 618 acres) is usual.

Ideally the propagation area for gladioli growing is changed each year, the area having gladioli growing one year in five. Different growers have their own ideas as to which crop is best to precede gladioli; in the USA land previously having soya or maize is preferred, whilst in Canada these plus tobacco are equally accepted. There is conflicting advice on this subject — some growers will even accept land on which cucumbers, melons or haricot beans have been grown. The previous-crop problem is never serious to the large-scale grower who takes into account the fact that in case of doubt the land will be pre-treated with a soil fumigant such as methylbromide or Bazamide, and the planting stock drilled in with Dysiston granules, a slow-release systemic insecticide which has proved to be extremely successful in all the gladiolus-growing countries.

Commercial propagators in the Netherlands, Canada and the USA set aside the required area of land on which they intend to propa-gate cultivars for the eventual sale of the corms only, thus ensuring no complications with cut-flower sales. In the Netherlands it is in fact illegal for a grower who is registered as a propagator to sell cut flowers from propagated crops. The only flowers allowed in the Dutch propagating fields are a few dozen in each block to verify that the cultivar is true to name and unmixed.

In order to be compatible with the use of field machines, the planting stock is raised in blocks or rows which are so spaced as to allow fairly precise machine cultivation during growth and at harvest. Three rows about 12 in (30 cm) apart about 100 m (yds) long, together with the outside paths constitute a block. If cormlets are used for propagation the cormlets are graded — only the larger cormlets are used of average size ¼–⅜ in (50–75 mm diameter).

The cormlets are soaked (or cracked) before planting, usually in a fungicide/insecticide dip. The drilling-in machine (seed planter) is set to deliver a cormlet each ¾in (2cm) along the row, which sounds very close together. Cormlets do not seem to mind which way up they land in the drill and given good conditions and a wet start the germination rates are good. Normal planting depth is from 2½ to 3½in (5 to 8cm) in a V-shaped drill.

By regulation the Dutch growers may use small corms too, and take regular supplies of what the Americans would call number 5 or 6 size corms. The average size of the corms is about 1 to 1.5cm (around ½in) diameter. Such size corms are not permitted in Holland or for export from Holland, to be sold as flowering corms.

It is normal after planting to use a weed pre-emergence spray to delay the weeds at least until the shoots of the gladioli are through. Normal irrigation, weeding and spraying techniques are carried out during the growing season by the same routine as for flowering crops. If flower stems appear on any propagating stock during August or early September these are cut off and destroyed, thus diverting the energies of the plant into corm production. At flowering time (or after deflowering) — normally 80–90 days after planting — the plants are given a side-dressing of a high potash fertilizer (10K-4P-4N) to improve corm size and density. It also induces fewer but larger cormlets at digging time. At about 120 days after planting (the planting date is invariably noted on the nameboard) the entire plant is cut off at ground level and within two or three days (subject to weather conditions) the entire plant, roots and all, is dug up and transported to the drying sheds. The plants are spread on mesh bottomed trays one corm deep and drying commences immediately. From then on the procedure is the same as for flowering-size corms.

Using this technique the Dutch growers produce quite a high proportion of large 14cm up, 12–14cm and 10–12cm corms of flowering quality. Smaller sizes 12–14cm, 10–12cm and even 8–10cm are more common with miniature and primulinus cultivars. Anything smaller than 8–10cm is not regarded as suitable for sale as flowering corms (though many are, unfortunately). The 8–10cm and 6–8cm are used as propagating stock. The smallest corms are included with the larger cormlets and used as propagation stock (sold by the litre!). Note that the size given refer to the *circumference* of the corm, and no imperial equivalents are given here as these are not applicable.

The entire Dutch operation is ranged around the soil and climatic conditions prevailing in the bulb-growing areas of the Netherlands. The greatest asset the Dutch have in this regard is the nature of the soil where gladioli are grown. The soil is either enriched dune sand (near the coast) or dredged soil from dykes and ditches in the reclaimed land (polder areas). The main growing areas are north and south Holland and Flevoland, and the reclaimed area of Friesland by Ijsselmeer (Zuider Zee of former days.)

The largest growers of gladioli corms in the USA and Canada have a quite different and even more difficult set of circumstances to contend with than the Dutch growers. They are however somewhat less restricted by state agricultural authorities than are the

Dutch. For example, there is no compulsion on the North American growers to destroy the flowers from propagating stock, nor do they have to choose whether they grow for corms, for flowers or for both. Basically the Canadians employ the same growing techniques as the Dutch and so also do those growers in the northern and midwest states of the USA. The reason for this is mainly due to climatic and soil conditions.

It is however not possible for Ontario, Manitoba, Alberta or the eastern Canadian provinces to leave their plants in the ground for 100–110 days without serious risk of frost damage. Nor can they plant as early as the Dutch. As a result of the short growing season Canadian and northern USA growers lift mostly smaller-size corms and have smaller cormlets. This is particularly true of the later-flowering cultivars.

A large grower in Michigan and another in Illinois overcame this problem by setting up special early-sprouting arrangements in warm sheds so that the late-flowering and expensive show cultivars are in and through just as early as possible. They also start off indoors large cormlets and small corms of these selected (and thus expensive) cultivars. In the Vancouver area of Canada they have a milder climate and a longer season, giving conditions quite similar to Europe, thus they can plant early and leave the lifting until well in to the autumn without the risk of frost damage. The propagation of the large late-flowering cultivars is very successful in the Vancouver area, and the corm sizes are invariably good with plenty of large cormlets.

The best propagation area for cultivars of gladioli of all types is almost certainly California, and there one finds a multi-million corm grower established over 60 years ago. At Oceanside, California, is the Frazee operation. In the lovely climate of California one can plant safely at almost any time of year. For the corms to be sent for growing in the northern states and Canada and for export to Europe, the Frazee propagation crop (small corms and cormlets of new cultivars) is planted Jan./Feb., lifted Aug./Sept. and dried naturally by sun heat. They produce some of the heaviest and healthiest corms available. The Frazees have a special requirement for huge jumbo corms of $3^1/_2$–$4^1/_2$in (9–11cm) diameter to send to Florida in October and to certain Central and South American tropical countries in November. It is interesting to see the Frazee Bulb Co. at Oceanside with its computer-programmed planting scheme. At any month of the year there are gladioli in bloom, crops being taken to the preparing sheds and corms being heat treated or cooled ready for planting.

In north-east Florida the Manatee Fruit Co. grow an enormous crop of winter-flowering gladioli. Using the huge jumbo corms they grow flowers for the Christmas and New Year market in the northern cities. The corms are planted about 3in (8cm) deep in long rows of up to 1,000yds (m) between the citrus trees in the Manatee Plantation. The corms are destroyed immediately after flowering. It is not possible in Florida to propagate cormlets successfully and economically, so the Floridans import the huge corms from California and secure very high prices for the flowers, which go with the oranges, lemons, grapefruit, etc., in temperature-controlled trucks to the north.

For reasons very similar to those quoted for California, the propagation of quality gladiolus corms in Australia is now being done in Queensland and Western Australia. The cultivars raised in Victoria and South Australia are propagated in the warmer areas with good reliable rainfall (usually) and, given the longer daylight hours and absence of frost, the corms and cormlets can be produced at the right time with bigger corms made possible by leaving the plants in longer. A top-size corm can be produced in Queensland from an average cormlet in 100–110 days. Australia is now in the process of developing an export business for New Zealand, Japan, Korea and the Philippines.

It is quite possible that by 1992, when the 'One Market Place Europe' becomes a fact, that a scheme similar to the Australian one will be launched with perhaps Dutch and British cultivars (or selected USA and Canadian) being grown in southern Italy and Spain in the winter months for spring/summer sale in Western Europe.

## Commercial Propagation Methods

### Meristem-Tissue Culture

A development in recent years in the multiplication and propagation of flowering plants has been the introduction of tissue culture or meristem cloning. The technique is so revolutionary it gave the perfectly harmless word 'clone' a sinister connotation that led science fiction writers to fantasise the application to humans, with the usual horror-film sci-fi endings. To reassure the nervous and to make tissue-cultured plants normal and acceptable, let it be said here and now that, at least at present, there is no sound scientific or medical evidence to support the view that human beings can be tissue cultured or cloned. The prospect of raising thousands of look-alike beings from human tissue is still, happily, a late-night TV movie scenario.

The basic theory behind meristem-tissue culture in plants is that it is possible, by taking a few living cells from a growth-generating part of a plant, and feeding the block of cells on a bed of nutrient jelly in the presence of ultra-violet rich light and a warm humid atmosphere, that the cell-block will rapidly develop roots, leaves etc., and become a viable plantlet like a seedling. The technique has been well established and documented. Given the necessary laboratory facilities it is not a difficult technique to master. In Florida, California, Japan, Singapore and Hawaii commercial units are creating literally millions of plantlets annually. Several genera are known to respond to tissue culture and these include orchids, lilium, cacti, African violet, ornamental banana and begonia. In Britain the John Innes Institute successfully pioneered tissue-cultured gladioli and freesia. About 8 million gladioli plantlets are produced by tissue culture annually in Florida.

Although tissue culture is a rapid and excellent way of increasing valuable specialised stock, it is an expensive operation reserved for high-cost plants or plants which cannot be propagated at an economic rate by the normal cultivation methods. Gladioli can be propagated at an immense rate by meristem culture. The donor corm can be cut up into microscopic cell-blocks of the right tissue and from one corm up to 1,000 small corms can be produced within about 90 days. The cultured plantlets do not of course go to flower; the aim is to produce a viable planting-size corm only.

The plantlets grow in a clinically perfect environment so that the increase and propagation can be continued all the year round irrespective of outside climate. The operation is done *in vitro*, as they say, though nowadays it is more likely to be clear acrylic plastic, sterilised as though the operation was medical. The agar jelly and special nutrient chemicals are expensive, and the equipment required to create the perfect microclimate for the plantlets is also costly. The tissue-culture unit is invariably computer controlled to ensure correct temperature and humidity, as well as the length and intensity of the artificial daylight with accurately devised ultra-violet light content. The entire operation is supervised by special bio-chemists or botanists – and they cost very much more than field labour.

A summary of the Florida tissue-culture project should be of interest to the non-scientific gladiolus enthusiast. The question as to whether tissue-culture techniques could be successfully applied in economic terms to commercial gladioli was first raised in Florida by the Manatee Fruit Company. The established traditional way of growing cut-flower gladioli in Florida was to use huge multi-headed 'jumbo' corms over $2\frac{1}{2}$in (6cm) in diameter. These big, flat corms had to be planted late August or early September to catch the flower markets at Christmas, and this time schedule meant that supplies of corms for Florida required special treatment. Suppliers in California and the mid-West harvest their annual crop in late September, clean and store it during the winter months and sell the mature corms in March and April for planting. For Florida supplies the growers had to select the very largest corms and retard their breaking of dormancy. This was done by keeping the large corms in darkened, cool storage in a partially de-oxygenated atmosphere (nitrogen blanketing) until August, then shipping them for immediate planting.

As a result of this prolonged and enforced dormancy many corms failed to germinate but became spongy and rotten in the ground; others grew, but were very prone to virus attack and 'fusarium yellows'. Where the virus and fusarium attacks were less severe corms flowered but the stems were short, and the flowers so crowded as to be virtually unsaleable. Following carefully monitored trials in Florida under the guidance of Dr Robert O. Magie and Dr Gary Wilfret, a few cultivars were selected from the commercially available stocks, of which those with resistance to fusarium and absence of visible virus symptons were set aside for further development work. Using the selected candidate cultivars in the programme, a series of corms were innoculated with the most prevalent and virulent strains of fusarium and virus. The cultivars that showed resistance to the innoculation treatment were then tissue cultured to give sufficient quantity of planting stock quickly. From this selection of cultivars there emerged two clonal cultivars that had high resistance and good flower quality.

The two cultivars of tissue-cultured origin have now been produced in the field in commercial quantity by usual propagation methods. The cultivars are listed as 'Dr R.O. Magie' and 'Florida Flame' and are described as follows:

'Dr R.O. Magie' (334) — Early to midseason; 75 days planting to flowering (NB in Florida). Medium-sized pink blooms with creamy-white throat. Straight, tall spikes, 20 buds with 8 open florets. Very healthy and productive.

The second cultivar is large flowered.

'Florida Flame' (454) — Midseason. 80–85 days. Bright red, strong colour. Tall tapered spikes of 20 buds with 8 or 9 open. Healthy and disease resistant.

The stock of these two cultivars now available commercially in quantity has been grown in many parts of Canada and USA with considerable successs on the florists' market. They appear to have adapted to field conditions remarkably well, and have adequately retained their disease resistance. Happily these cultivars are stated to transmit the healthy characteristics to seedling offspring irrespective of whether the pollen of the disease resistant cultivar is used or the seed is carried on it.

Corms retrieved from tissue-culture specimens are high crowned, almost conical, with a circumference of 5–6 cm (2–2½in). These corms are delivered dry and dormant and are planted and cultivated in the normal way. As would be expected there is no 100 per cent guarantee that all corms will be disease-resistant, especially if planted on ground that has previously grown gladioli, freesia or any other member of the iris family (Iridacae). Ruthless culling, with incineration of affected plants, is imperative for these tissue-cultured stocks. Reports indicate that in Florida the rate of culling (or rogueing as they say in Florida) is not more than 5 per cent per crop cycle.

Other cultivars in different colours are now being tissue cultured — most of them being seedlings from the virus-free parents. Some seedlings from 'R.O. Magie' and 'Florida Flame' are currently on trial in USA and the originators are confident that this new generation of cultivars will be as good as their parents.

The viability of tissue culture as a means of virus-freeing gladiolus cultivars is now accepted, though perhaps the economics are still in question. The next step is to use tissue culture as a means of producing in a short time plentiful stocks of small corms of new cultivars that have shown promise as seedlings. Some of the new cultivars bred for exhibition purposes and others of more exotic, fancy or unusual form are very slow propagators, seldom producing cormlets or dividing the mother corm. In these circumstances it is difficult for a raiser of a new gladiolus cultivar, however, desirable or prestigious it may be, quickly to produce the number of corms needed for trials, or to have it seen on the show benches nation wide. Here tissue culture could be of assistance. It is estimated that tissue culture can easily multiply by a factor of 1,000 the amount of planting-stock corms that one corm could produce in two seasons.

Tissue culture is likely to provide the answer to rapid increase in availability of Grand Champion show stock. Only by this method could an exhibitor be guaranteed that he is getting corms from the actual Grand Champion, and what a price that 'block of cells' from 'the best gladiolus in the world' could command! That one corm could be just as valuable as the sperm bank from a champion Charolais bull. Now we're back full circle to 'animal clones' — or are we?

### References
G. Hussey and Judith Hutton: *In Vitro* Propagation of Gladiolus by Precocious Axillary Shoot Formation. John Innes Institute *Annual Report* 1965-67.

*The Gladiolus Annual 1978*, British Gladiolus Society, pp 38–40: '90,000 cormlets from one corm in 18 months'.

# *Chapter 4*  **Cultivation**

The cultivation procedures for the *Gladiolus* species, the species hybrids and similar small-flowered gladiolus cultivars will be dealt with in the relevant chapters where these types of gladioli are described. The special techniques that have been developed for the cultivation of exhibition types of summer-flowering gladiolus culti-vars will be fully described in the section dealing with growing for exhibition. This chapter will concentrate on the more general culti-vation methods as applied to the summer-flowering 'grandiflora' type of gladiolus as grown for cut-flowers, landscape, garden decor-ation or floral arrangements.

The garden gladiolus does not produce a bulb but a corm, which is a globular compressed rhizome or underground stem. This state-ment is made not as a botanical phrase of one-upmanship, but to indicate that being a corm, the size of the starting unit is of less significance than is the case with a true bulbous plant. In a bulbous plant the embryonic flower is contained within the bulb. In a corm the embryo flower is formed later inside the shoot. Provided there-fore a corm of gladiolus is healthy, with sound root nodules and a developing shoot it can, with good cultivation, be grown to develop a flower from quite small corms, as small as ½in (1.2cm) in diameter. The best sizes to plant for normal garden use however are top-size 14cm up (the circumference measurement), 12–14cm or 10–12cm for large flowered; the medium and miniature-flowered cultivars seldom produce the larger 14cm corms. The huge flat 'jumbo' corms of 16–18cm circumference are not recommended as they are old and have lost their vigour. By reference to the cata-logues of suppliers and using the Classification Colour Code as guidance it should be possible to select gladiolus corms of the described colour, size and season of flowering with reasonable ease. It is well worth paying a little extra to obtain supplies from reputable and established gladioli specialists.

Gladioli will grow on allotment soils, and a little 'tender loving care' and attention to detail in soil preparation and planting will reward both the gardener and allotment holder with far better flower spikes. All perennial weeds, large stones and heavy clods of soil should be removed before soil preparation begins. If the soil is low in organic food then peat, garden compost or well-rotted manure could, with benefit, be added to the lower levels of the soil. The roots of gladioli seldom travel further than a depth of about 8–10in (19–24cm). For this reason a balanced fertiliser, bone meal or fish manure is best incorporated in the layer of soil that will be ultimately 6–8in (14–19cm) below the surface. The entire planting area should be well forked, rotovated or spaded to a depth of about 12in (28cm), with the top 6in (14cm) made loose immediately before planting. Choose a site which is in full sun from April to September with, if possible, a windbreak fence or hedge on the side facing the prevailing wind.

Planting patterns i.e. rows, blocks, clumps etc., will be a matter of personal choice according to the end use. Planting depths will be the same irrespective of use, namely: large-flowered cultivars 6 in deep (14 cm) and 10 in (24 cm) apart; medium-flowered and Butter-fly types 4 in (10 cm) deep and 8 in (19 cm) apart; primulinus and miniatures need about 4 in (10 cm) depth of soil above the top of the corm to support the flower stem, as at this depth canes or supporters will not be necessary. These smaller-flowered varieties can be 5 in (12 cm) apart with added effect. Unless large quantites are being used corms are best planted individually by hand trowel. Alternatively, a round-ended dibber can be used on sandy or friable soil but never on clay or heavy loam. If your soil is of clay or heavy loam, or if the subsoil occurs at a depth of 6 in (14 cm), the hole into which the corm is to be placed should be lined with a mixture of equal parts of sand and peat. This will assist drainage and ensure a clean lift in the autumn.

Immediately after planting it is often useful to apply a pre-emergence weed-inhibitor spray. This will retard the germination of annual weed seeds so that weeding or hoeing will not be necessary until after the gladioli shoots have emerged and are clearly visible. If cormlets or seed of gladioli have been planted it is better not to apply pre-emergence weed sprays on these plantings. Whilst it may not kill the seedlings or cormlets, it could cause leaf distortion which could impair growth.

If soil conditions at planting time, in April or early May, are dry it is advisable to soak the corms overnight in clean rainwater to encourage rootlet growth. At this stage the addition of Benlate in liquid form is helpful in making the dip fungicidal, or Jeyes Fluid at 5 ml per 5 litres water (1 part per thousand by volume) is a good alternative for treating large corms. Unless drought conditions prevail, irrigation is not usually necessary until the growth-shoots emerge. Gladioli should never be allowed to dry out at any stage during the growing period. In very dry periods a generous flooding once a week is far better than daily sprinklings. Cormlets should be in moist soil at all stages to ensure that good sized corms may be harvested. At the five-leaf stage a weak solution of a fertiliser like Phostrogen, Baby Bio etc. may be used every 10 days, but make sure the formulation has a high potash content, and avoid high nitrogen feeds.

A preventative insecticidal spray should be lavishly applied in early June to cater for possible thrip attack and to kill any caterpillars, aphids, or frog hoppers that may arrive. In the event that signs of thrip attack are seen a second spray just as the flower stems are forming sees them off. Thrips are best dealt with using a systemic insecticide at 7-day intervals until the infestation is clear. The eggs of the thrip insect are not killed by contact insecticides, but systemics properly and carefully applied will kill the young thrip larvae immediately they hatch. Any proprietary systemic insecticide will be effective, though those formulated with Dimethoate seem better suited to gladioli.

During the growing period beds must be kept weed free, as not only do weeds use up the moisture and feed to the detriment of the gladioli, but many weeds act as host to the more prevalent insect pests such as white-fly and aphids. In the last few days of July, or a little earlier in the southern parts of Britain, it is useful to give an additional liquid feed of sulphate of potash (potassium sulphate). This will help to build strong spikes, longer-lasting flowers and will improve flower colour in darker gladioli. A rate of 50gm (2oz) per gallon of water (approx 10gm per litre) is a good average dosage. Be careful not to overdo the potash as it may cause white, cream or pale-yellow flowers to produce red flecks in the petals.

Some of the heavy-leaved and thick-stemmed gladioli (particularly the large-flowered American and Canadian cultivars) have difficulty in forcing the flower stem through the leaf sheaths. The leaves exert strong pressure on the flower stem which will crook when it gets trapped in the sheath. To prevent it becoming bent and distorted it will be necessary to open up the leaves with a thumb nail and gently manipulate the tip of the stem out of the sheath. The stem is fairly flexible at this stage and by careful tying to a support it can be straightened. The release is best done in good daylight whilst the temperature is up. A good watering with cool rainwater will restore the rigidity of the stem.

To obtain the best from gladioli as cut flowers it is recommended to cut them early in the morning when the first floret has just opened. More details about this and advice on how to prolong the life of the flowers after cutting are given in Chapter 5.

Most of the available garden cultivars of gadiolus flower between the end of July and the first week of September. There are however a few very late cultivars that may not flower until mid-September or later. Unless the flowers are needed at this late stage it is best to avoid the late cultivars for two reasons. First, the weather in September is or can be cooler and wetter than August and the daylight lengths shorter. Under these conditions the flowers take much longer to open and are also more prone to petal rot and markings due to acid rain. Secondly, and importantly if you pay a lot for the corms, the size of the corm that can be dug up is often smaller than the original planted corm, and is without cormlets. This characteristic of late cultivars causes propagation problems, the unripe corms being difficult to store over the winter period.

Once the flower stem has been cut off the energy of the gladiolus plant is diverted to the production of new corms and cormlets. To achieve this the leaves of the plant must remain green and active for about 40 days (6 weeks). Continue during the 40 days to treat the plant as you did pre-flowering. Keep weed free as this will reduce the risk of the spread of fungal disease, like damping-off or botrytis. Reduce the watering to half the usual frequency, or if rain intervenes, no irrigation at all. At approximately 40 days after flowering the corms should be mature and cormlets will have formed. Cut off the entire foliage to about 2in (5cm) above soil level. Any plants from cormlets or small corms should be cut at the end of September whether they have flowered or not.

Lift the corms, using the 2in (5cm)-stem as a handle, and place them complete with cormlets in a seed tray. Choose a dry, sunny day, if at all possible, or at least when the soil is dry. At lifting stage any plants with dry, brown, yellow or withered leaves should be trimmed right to the upper corm and examined carefully for scabs or rot. Any corms which appear defective or unhealthy should be stored and dried separately. Trim off the roots and the top and place the lifted corm as quickly as possible in a well-ventilated location at a temperature of about 25°C (70–80°F). An electric fan heater or an open airing cupboard are most suitable as adjuncts to good drying procedures.

After about 10 days any soil will have dried and can easily be shaken or scraped off the corm. A quick twist should easily detach the old corm and the clusters of cormlets (spawn). Clean up the new corm to leave a clean root scar, and detach the remnants of the old stem from the centre of the upper part of the corm. If you wish to propagate any particular cultivar select about 20 to 30 of the largest cormlets and store them safely in small bags or envelopes (not plastic). This simple precaution will prevent the loss of valuable cormlets. It is so easy, when cormlets are stored with large corms, to lose large cormlets from the bags, since the bags are left open for air circulation. Plastic nets tied with nylon twine, or discarded nylon stockings (or half tights), make excellent storage containers. These also have the added advantage that they can be used when corms and cormlets need to be dipped before planting (in fungicide or insecticide).

Before placing the clean, dried corms in winter storage dust them with insecticide and fungicide, 50/50 green sulphur and Gammexane is effective. Store them in a dry, airy cupboard or store-room at an even temperature around 50°F (10°C). Inspect your stock at intervals during the winter and check that there is no damp-ness or condensation in the store. Any corms that have shrivelled, gone stone-hard or spongy–rotten should be taken out and burnt; there is no way these corms will ever produce good plants. Use a new sack for the undamaged corms of the same cultivar and re-apply fungicide dust.

As planting time approaches — early March is a good time — take the corms out of storage and look at a few to see if they have begun to develop shoots and root nodules. If they have they can then be 'started'. This means the corms should be brought into the daylight and stood shoot uppermost in seed trays or flat boxes. Make sure the light is good and coming from above, otherwise the shoots will develop at an angle facing the light source and will elongate too much. Temperatures of around 55–60°F (12–15°C) will induce germination and dormancy break, giving the corm a flying start with root nodules at the ready and sturdy green shoots upright and pointed.

Should you decide to grow on your own flowering-size corms from cormlets or small corms it is well worth de-husking them prior to planting. This is necessary because cormlets and small corms do not readily develop shoots and root nodules whilst still encased in their husks. A pre-soak in tepid, clean water before planting — overnight will do — assists germination. Cormlets need not be planted upright as they will find their own way through the ground from any position. Small corms however need firmly placing upright to ensure good germination. For this reason it is advisable to line the holes or rows with a mixture of sharp sand and peat and press the small corms into this medium before covering over carefully with earth from which stones and lumps of soil have been removed.

If you are propagating expensive or favourite cultivars from cormlets or small corms label them carefully using permanent marking ink, and also make a plan in your record book. There is nothing more frustrating than having a good lift of a cultivar and not knowing what it is. Without a good label, if it does not flower you may never know what it is.

A final word of warning: do not be tempted to use selective weedkillers or pre-emergence sprays on valuable cormlets or small corms. It may damage these small and tender growths or at least bring about leaf-distortion problems which may easily carry over into the harvested stock. Accidental damage due to contamination by lawn fertiliser is not uncommon, and occasionally from straw originating in cereals from fields dressed with 2-4-D or similiar selective herbicides.

# Chapter 5   Gladioli in the Garden and the Home

Twenty-five to thirty years ago it was quite common for gardeners to say that they did not get good value for money when growing the gladiolus cultivars then available. An oft quoted reason, and a very practical one, was that gladioli flowered mostly in August when the grower was on holiday, thus missing the pleasure of seeing the result of the season's work. Another reason quoted was that gladioli were too stiff and formal, often too big to be appropriate in home decoration, and did not last very long in arrangements once cut.

In the 1960s those criticisms of the gladioli were to a large extent true and valid. Commercial growers depended on mass sales of easily grown self-coloured cultivars. To the large grower a cultivar that could be relied upon to produce stiff stems of even length at a fairly predictable date was the ideal bloom. Alongside the 'commercials' were a much smaller number of large-flowered, very tall and formal cultivars for exhibition purposes.

It was the fashion of the time in the 1960–70 period to regard only as worthwhile gladioli those cultivars which by virtue of their ability to win Grand Champion at national shows achieved fame and publicity not only for the exhibitor but also for the raiser. These huge, many-budded gladiolus cultivars were exclusively exhibitor material, and in the event they opened the florets too early or too late for the show in question and were not all suitable for use in the home, being close to 6 ft (1.9 m) tall with a 2 in (5 cm) diameter stem.

Times have changed, and much progress has been made in all the gladiolus-growing countries. The exhibitor, gardener and floral artist now have an excellent and comprehensive range of colour, size, type and season of flowering. The versatility of the gladiolus flower is firmly established.

## Use of Giant and Large-flowered Gladioli

The large and giant-flowered cultivars still have a role to play in the display of the gladiolus. In places like hotel foyers, churches, banqueting halls, reception areas in large buildings and the like, a massive, fan-shaped arrangement of large gladioli is an imposing sight. The big one can, in moderation, be used occasionally in the home — a fireplace can be 'lit up' in the summer with an earthenware jug full of brightly coloured red, yellow or blotched gladioli. A few large-flowered gladioli in a suitable container at the foot or head of a staircase brightens up a hallway and provides a focal point or diplomatic distraction. Large white gladioli are best reserved for church decoration where they are very appropriate. There is not a lot of opportunity to use white gladioli *en masse*. There is one near certainty about large-flowered, white gladioli: they are far from ideal as garden flowers. Apart from their limited use it costs a fortune in stakes or canes to keep them upright at flowering times.

Post-war houses and gardens, having rather smaller areas to exploit the gladiolus flowers, have created a demand for smaller-flowered gladioli for floral arrangements in the home, and shorter, robust, medium-flowered cultivars for the garden. The former are called 'miniature gladioli' and the latter 'landscape gladioli'. The important requirement of the landscape gladiolus is that it can hold a full-size, fully opened flower stem upright without stakes or cane support.

## Landscape and Garden Gladioli

Landscaping is really garden decoration on a grandiose scale, so the same considerations apply to both. Don't be tempted to buy more corms than you can properly handle just because 250 corms of one cultivar are so much cheaper than 50 each of 5 cultivars. There are two major considerations, bearing in mind the flowers will not be cut but will remain in the garden until they completely fade.

For the best effect the flowers of each cultivar must come at more or less the same time; for garden use batches of 10–12 corms per cultivar is ideal. Choose a selection of early, mid-season and late-flowering cultivars if you want a long spread of season from late July to early September. In general the early flowering bloom in early August, the mid-season from mid-August to end August, and the lates from end August to mid-September. Those who are likely to be away in August are advised to plant earlies or lates only. Colours planted side by side should harmonise as there is bound to be some overlap, especially if the corms are of variable sizes.

The second consideration is that the chosen cultivars must have strong, shortish stems and a neat, balanced flower-head. The large-flowered cultivars when in full bloom will become top heavy and topple over, especially in rain; and with canes or supporters costing so much, and also being an unsightly intrusion in the garden, it is best not to have to use them at all. Choose the medium (300 size)-flowered, the miniature (200 size) or primulinus. Avoid the giant (500 size) at all costs. If there is a sheltered part of the garden some 400 size (large flowered) may well grow unsupported.

Plant the corms at the appropriate time according to flowering date required, about 5 in (12 cm) deep and 6 in (15 cm) apart in groups or circles rather than straight lines. A good format is a tri-angular group 4-3-2-1 (for 10 corms) or 5-4-3-2-1- (for 15 corms). Alternatively, arrange the corms as on a clockface of 15 in (36 cm) diameter, at 'hourly intervals' and three in the middle. It is better and more effective to have a few cultivars of 15–20 corms per cultivar than to have fewer corms of many colours or shades.

To prolong bloom life remove dead florets each two days and do not let seed pods develop. It is not necessary, as is done with exhibition cultivars, to remove the side spikes. The secondary spikes will lengthen the colourful period and will not impair the corm. If two or more shoots arise from one corm do not remove them — let them all grow. An extra feed of high-potash liquid feed just before

*Gladioli in the
Garden and the
Home*

the flowers show colour will help the plant to cope with its multi-flowered effort. A suitable high-potash feed liquid is Tomorite or Compure K; the solid form can be made by adding 1 kg (2.2 lb) of potassium sulphate to 3 kg (6.6 lb) of Growmore fertilizer.

Strong self-colours are most effective *en masse*, but a contrast can be made by using an occasional blotched or Butterfly cultivar. It is best to avoid the very dark colours such as black-red, deep purple, or deep violet, and limited use should be made of grey, smokey or brown shades in the garden as these appear to be prematurely finished when viewed at a distance. (Though these 'off-beat' shades can be grown as cut flowers in the kitchen garden). Dark shades tend to spot or fleck badly if subjected to thundery rain.

In sites that are exposed to wind it is worth considering the primulinus hybrids, now available in a wide range of colours. The strong, whippy stems of primulinus gladioli will yield to the wind without suffering damage, but remember they may be 10 in (24 cm) shorter than the other (300 size) cultivars and up to 14 days earlier to flower. If you use primulinus delay planting until at least two weeks after the grandiflorus types. A list of recommended landscape, garden and floral art cultivars is given in Chapter 9.

Prices and availability will dictate your final choice, but avoid the pitfall of growing mixtures because they are cheaper. You will end up feeling disappointed despite having saved money.

## Home Decoration

A supply of suitable gladioli for home decoration can be provided from 'the glad patch', an area set aside exclusively for the growing of gladioli for cutting or for exhibition. A number of cultivars suitable for home decoration is given in Chapter 9; these cultivars are recommended for growing to produce arrangement gladioli, and the arranger may be able to add a few of the exhibition miniatures which are flowering at the wrong time for the show.

I shall not presume to advise or suggest what you do with your gladioli when making your arrangements. That is a very individual thing, and I would no more think of telling a gladiolus enthusiast how to arrange gladioli than I would tell him what wines he ought to drink. What I do intend is to help you make the most of your gladiolus material.

The modern gladiolus has far better keeping qualities than those of 20 years ago. An arrangement properly made and using an Oasis foam will last several days without deterioration, but there are one or two important provisos to ensure long-lasting arrangements or vase displays. Cutting time is important, especially with the frilled and ruffled miniatures. Without doubt the best time to cut gladioli is early in the morning from well-watered plants. Cut the stems low down with as few leaves as possible when the first two (i.e. lower) florets are just about open. The cuts should be made with a sharp knife or secateurs to give an oblique end to the stem — this exposes a larger area of the interior of the stem thus enabling the stem to take up more water. Harden the stem by placing the flower spike in deep, cool water to which has been added some Chrysal or Baby

Bio. If the flowers have suffered in the heat of the day the addition of cane sugar (4 per cent by volume) will help the florets to recover and firm up. If the flowers are far enough open already, keep them in a cool, well-ventilated dark room. If on the other hand they need more opening a warm room well lighted will be helpful. Before arranging in the final display push back the green bracts on the upper buds to expose that little hint of extra colour on the tips. Better balance can often be created by nipping out the three buds at the tip.

If the middle florets are not open far enough to create the desirable taper on the stems push your little finger gently into the tube formed by the petals and gently ease them open. The florets will remain as far open as they have been pushed; if they move at all it will be to expand rather than contract. If undue force is avoided and only fingers are used, the technique is quite safe on properly prepared stems, as described above. At the final stage, particularly in vase displays, make sure you do not have leaves (other than gladiolus or iris) below the water line as these will decay rapidly and set up slime formation.

Often arrangers desire to have gladioli with stems that are not straight (i.e. Japanese-style arrangements); if so the best rule is to get a straight one and bend it to the required profile. This is done by leaving the flower out of water, flat on a table in a warm room for 4 to 6 hours until it is totally flopped. Take a length of strong but flexible wire (plastic-coated fencing wire is most suitable) and bend it to the required shape. Tape or twist-it ties can be used to secure the stem to the wire using a tie at intervals of about 2in (5cm) along the stem. If necessary use extra ties on the curves. Once the stem is bent and secured in the desired shape, re-immerse the stem end in cold water and it will set rigid and so remain. Remove the wire and ties carefully immediately before making the arrangement. Leaves can be shaped into curls and spirals in a similar way. If plastic-covered wire is not to hand strong basket cane may be used as an alternative.

If you find that certain stems of gladioli do not have regularly placed florets and this defect could affect the balance of your arrangement, it is quite easy to attach extra florets to fill the gaps — but do be sure you use florets of the same cultivar and colour. It is most important that the floret to be attached is carefully cut from the stem with a sliver of stem still attached to the calyx. This sliver of stem should include tissue from above and below the bud junction. By carefully taping the floret in position using green plastic twist-it or florists' tape an almost imperceptible 'fit-in' improvement can be made. Never wrap wire round a floret-tube as wire will cut into the floret and cause it to wilt prematurely.

## Corsage

Most attractive corsages can be made using the small and miniature-flowered gladioli either as single florets or tips of gladioli with two or three buds still fresh. Strip off any faded florets then take either a single floret with about 1.5in (4cm) of stem attached, or two small florets plus an unopened bud if needed with 2in (4.5cm) of stem still attached. Tape in some fern, asparagus leaf, maidenhair or other suitable foliage and complete the piece with a fine wrap of aluminium foil around the entire bottom, including the stems of the gladioli and foliage. If the greenish-lilac and pale-coloured, orchid-flowered Butterfly or primulinus hybrid gladioli are used, the effect when the corsage is worn upside down is like that of an expensive orchid.

## Posies and Sprays

On a slightly larger scale but still using the small, miniature-flowered or primulinus hybrids some delightful mini-bouquets, posies or sprays can be created using the same technique as for corsage. Cream, yellow, lilac, rose and lavender-coloured gladioli or the greenish-yellow, orchid-flowered hybrids are particularly effective in bridesmaids' posies or sprays. Used in conjunction with freesia, alstromeria, small lilies and trailing foliage, these sprays can be quite outstandingly different and create lively interest.

Now that the ruffled miniature, frilled, small-flowered and exotic gladioli have become more readily available, small hand bouquets and sprays are receiving much more attention from florists for 'special occasion' floral tributes at receptions, wedding anniversaries, prize-givings etc. Suitable-sized gladioli are available from June to September in a variety of novelty types, colours and forms, and these new gladioli can be coupled harmoniously with a large range of other summer flowers. If the complementary flowers are fragrant that is a distinct advantage. Freesia and bud carnations or pinks are particularly suitable.

The material for sprays, corsages, posies and other similar floral creations based on gladioli can all be grown under reasonable garden conditions — no expensive greenhouses are needed to cultivate the components. This results in expensive-looking creations at economic cost, and the recipient scarcely ever realises that you haven't paid pounds for the offering.

## Use of Foliage and Complementary Flowers

Most arrangements where gladioli predominate can be improved by the use of the foliage of other plants and the juxtaposition of complementary flowers. To complement the gladiolus foliage the leaves of *Iris, Montbretia, Crocosmia, Hosta, Kniphofia* (red hot poker), *Typha* (bullrush), maize or pampas grass are admirable. Some of these plants also produce variegated forms with either cream or yellow stripes. The variegated forms are especially effective when cream or yellow gladioli are included in the arrangement.

Foliage of a feathery or fluffy nature can be used to give the impression of 'bulk' to a gladiolus arrangement. *Alchemilla mollis* (Lady's Mantle), a wild flower of moist places, is light lime-green both in flower and foliage. *Achillea* (Yarrow) has dark-green, feathery leaves and makes an excellent foil. The flowers may also be left on if the white, yellow or cerise colours are harmonious. Fennel and rue are herbs with attractive foliage; Artemesia has attractive silvery foliage, and *Senecio Maritima* (Silver Dust) is similar but has more silvery and finely cut leaves. *Gypsophilla, Francoa, Nigella, Escallonia* and *Tamarix* are suitable as background material. They all have finely cut foliage and very small flowers giving the 'graceful ethereal' look.

More robust foliage in blue, green or variegation can be added judiciously to accentuate or contrast with gladioli. Better subjects in this group are; *Acer* (maples), especially in the autumn when the colour has developed, and a perfect foil to red, brown and dusky gladioli. *Eryngium* (sea holly) has deep blue, grey, spikey leaves with a silvery effect. *Fatsia* (figleaf plant) can be green or variegated. *Thalictrum* (meadow rue) has blue-grey leaves. *Pelargonium* (*Geranium*) has variegated and zonal leaves of green, white, yellow, red and maroon rings on the leaves. Remove flowers.

Flowers of a similar shape to the gladiolus make a very effective adjunct to a gladiolus arrangement. Amongst the most effective are *Freesia, Alstromeria, Schizostylus, Hemerocallis* (day lily) and some of the smaller hybrids and species of Lilium. Iris and Canna lily may be appropriate for very large vase displays. Annual godetia (dwarf variety) is also effective.

Composite flowers (daisy-like) and similarly shaped blooms can be used as a centre-piece effect from which the gladioli stems will appear to emerge. These include calendula, cardamine thistle, coreopsis, cosmea, dahlia, globe artichoke, gaillardia, rudbekia, scabiosa and zinnia. The suitability of the larger composite flowers such as globe artichoke and thistles will be determined by the overall size of the arrangement. The smaller-flowered composites, such as coreopsis, are useful in pedestal arrangements since they 'fill' without adding any sense of 'weight'.

An important feature of the complementary flowers and foliage recommended is that they are easily grown in the garden and can enhance the garden decoration or landscape gladioli *in situ*.

# *Chapter 6*    **Growing for Exhibition**

Many gladiolus growers cannot resist exhibiting their blooms, the product of a full season's loving care and attention, the pinnacle of the growing year. They enjoy the hustle and bustle of the show, the friendly rivalry and social competition, a chance for enthusiasts to meet and discuss their favourite flower, to discover new cultivars, to see the results of the ultimate in cultivation techniques and the precise attention to detail necessary in the preparation and presentation of the exhibits.

There is a particular atmosphere about a flower show, especially under canvas and on grass; the characteristic scents linger in the memory to comfort during the long winter months and in anticipation of another season. Gardeners, forever looking forward to better things, are born optimists. Above all, it is the relaxation from everyday problems that life seems to impart unmercifully upon us that we find at a show, and for a short time at least we are at peace with the world. What can be better than to spend that time in the presence of fellow enthusiasts and in the wonderment of flowers?

## Initial Planning

Exhibiting gladioli is not easy: it requires determination and dedication, and there are many disappointments which make the successes even more rewarding. If it was easy the achievements would not be worthwhile.

Having once decided to exhibit, perhaps being fired with enthusiasm seeing these magnificent spikes at a major show, you should start to plan immediately. If you are to be successful all the basic and essential tasks should be carried out in the correct manner and at the correct time. I will guide you through the full programme, and can assure you that there are no secrets that will give easy and instant success. Over the years I have tried all kinds of ideas and experiments; most have been discarded as of no or little value, or in fact have caused more problems than they are worth. I repeat: to be successful the basic and essential tasks should be carried out in the correct manner at the correct time. The only other ingredients are just plain common sense, enthusiasm, determination and dedication.

## Selection of Cultivars

Some cultivars would never win at a show in a million years. Some might just do so on occasion. There are some that win regularly, and these are the ones to grow. They probably have been bred for just that purpose and from generations of cultivars that have also been winners. Your entire stock should consist of these bankers and ten different cultivars is about the maximum, indeed less may be better. Remember that it is also necessary to match spikes in some multi-spike classes, which is easier than having a hotchpotch of very many cultivars.

There is a temptation that is difficult to resist, and that is the fascination of new cultivars having glowing descriptions in the catalogues. Do not believe all that you read. Some of the authors go into wonderful raptures regarding the new releases, but I am afraid in many instances their literary talents far outweigh their gladioli talents – or indeed their honesty. Under the Trades Description Act they had better watch out in future, except that they always seem to have in small print somewhere within the catalogue 'we cannot give any warranty, expressed or implied regarding the performance of any cultivar since the customer's growing conditions are beyond our control.' Better let others spend their money on new, untried releases; you have not the time or space to waste when growing for exhibition. There are enough problems as it is; try to limit them. But do keep your eyes open at the shows and recognise the ones that are starting to win and have potential and show consistency. Research can be made of the successful cultivars in say the last three seasons at the major shows in the British Isles. *The Gladiolus Annual* published by the British Gladiolus Society will provide this information. Disregard the compiled lists of the top ten in America and Canada, some but not all may be also successful in the British Isles, but remember that the climates and conditions are far different and it is our own country in which we are growing them. Certain very successful cultivars from overseas have performed less well in England and especially late-flowering ones that have not the chance to reach their full potential in our usually cold and wet autumn. Keep your eye on the tables but keep your chequebook in your pocket!

The odds can be reduced to give increased chance if it is decided to specialise in one type, be it large-flowered, miniature or primulinus. I suggest furthermore that the selected banker cultivars include those from Holland, USA and Canada and also early, mid-season and late flowering. The catalogue should give the details. Obviously we don't want all the plants in the glad patch to be in bloom all at once, but spread over a few weeks to cover the different dates of the main show events. Even so, this is not easy, and we have no control over the weather. We could be lucky and have our best 'cut' that coincides with the National Show and scoop most of the major trophies. This has been known to be the situation on isolated occasions, but it is far better to spread the odds by permutation.

After some years' experience an exhibitor will have his/her own favourite bankers, those that consistently produce prize-winning spikes and respond to the particular conditions provided – the old theory of 'horses for courses'. Successful exhibitors also become associated with certain successful cultivars, having discovered the method under which a particular cultivar performs best.

## How Many to Grow?

To grow glads to exhibition standard requires much care and attention throughout the growing season, but the most demanding time is the few short weeks and days before the shows. If too many are grown then the required individual attention will be neglected and the quality will suffer, the grower will suffer through overwork and enthusiasm soon wane.

I cannot give you the precise recommended number as this depends on certain factors and personal considerations; but it is my opinion that no one without assistance, even if you are retired or are fortunate enough to have plenty of time, can grow many more than 1,000 corms for amateur competition. Many experienced and successful exhibitors at the national shows grow a lot less, and most at this level grow between 450 and 1,000 — but there are exceptions! It has been known for major awards to be won by a person growing a couple of hundred or so, but perhaps even after undeniably growing them very well indeed that person has been rather lucky to have a fair quantity of spikes to choose from on the day.

If local shows are the main objective and there are a number within easy reach of your home and they cover a few consecutive weekends, perhaps other subjects also being grown, then maybe 250 or even a few less is about the right number. It all depends on circumstances, but I have to be honest and say here and now that not every corm grown, even if grown really well, makes it to the show table. There are a lot of spikes that are just not quite right either in timing or in other respects. We all have some super spikes on the Wednesday when the show is on Saturday for instance. All is not lost however, and the beautiful blooms can find their way into a vase in the house or can make marvellous gifts to friends and neighbours or go for church decorations.

The most I have ever grown for exhibition is 1,250, and that alongside my seedlings and propagation stocks proved far too many. This was reduced to 750, which was found to be more manageable, the quality improved and enabled me to win top honours at national shows, this being my most successful period. Again it all depends: after all, the growing of gladioli for exhibition should be for pleasure and relaxation and not be regarded as hard work.

## Where to Grow

One is very lucky indeed who has unlimited and suitable land available and suitably located. Undoubtedly the best is a grazed, virgin field that can be prepared in the autumn for planting in the following spring. The balance of nutrients and the pH will probably be correct or require little adjustment and the fibrous roots of the turf will make the structure and the texture just right, especially if the basic soil is also light and well drained. There is no doubt that the organic content will be at the maximum if the land has been down to field and grazed for a number of years. There is one important precaution though: wireworm is often present in such fields, so do use BHC dust at planting time.

Glads like full light and sunshine, but if there are also hedges and walls some distance away that do not impede the rays of the sun but provide protection from the prevailing winds, then so much the better. Avoid tall trees that cast shadows. At the end of the season grass down your growing patch and move on to the next virgin patch.

That is the ideal — we can all dream can't we? — and most of us grow our exhibition gladioli in far from ideal conditions and locations. We can only do the best under the circumstances available to us.

Perhaps you have a fairly large garden attached to your home — there is no doubt that this is best for convenience when the plants demand care and attention and preparation, especially at flowering time. The few minutes before going to work, or if you are lucky, at lunchtime, or on return home in the evenings are all very precious. But glads do not like to be planted in the same soil for more than two years, and it may not be possible to rotate them with other crops to avoid this. Digging up the lawn may be the only resort! Renting an allotment may be the answer, especially if the travelling distance is not too great. Having water available is a necessity, however and vandalism could be a problem, but many exhibitors produce fine blooms from their allotments. Renting some land from a farmer or other landowner may be a further possibility. One of the most successful exhibitors on the national circuit at present rents three growing plots at different locations in his village. He always has some spikes available to cover most of the show dates.

Still in consideration of laying off our bets and to give us more chances, it is a fact that the environment, temperature and climate can vary dramatically in only a very small distance; even the micro-climates can be different and change within the location of a small garden. In such a case it will be an advantage to have two or even three glad patches. These different locations can have slightly different flowering times and therefore spread the chance of having some spikes at least just right for that important show.

As I said earlier, we can all dream of the ideal but it is surprising what can be achieved with enthusiasm and determination and a friendly approach. Perhaps there may be a large, unkempt walled garden just up the road which the owner may not now be in a position to cultivate due to age or infirmity. You never know unless you ask; the occupier just could have been a very keen gardener at some time previous, maybe a gladiolus grower! A trim of the front lawn once a week and the bargain is struck.

## Preparation of the Land

Once you have decided where to grow your gladioli for exhibition then preparation of the soil is the next essential phase of the programme. This should commence in the autumn and before the soil is too cold and wet to work satisfactorily. More harm than good will be done if the conditions are not ideal, so leave well alone. The ground should be cultivated by spade or mechanical cultivator depending on the area to be done and the facilities at your disposal. I prefer the spade, over which there is more control — and a little healthy exercise on a nice autumn day is good as long as not over-done. I have found that deep digging is not necessary and about 10in (25cm) is satisfactory. Yes, I have grown good glads by deep trenching but certainly no better. A problem could be if there is a layer of clay or heavy soil under the top spit; better not disturb and ruin the composition of the top soil unless of course the drainage is so bad that the clay just has to be broken up. You are lucky indeed if the basic structure of the soil is gritty loam.

Again, good glads can and have been grown by using vast quantities of humus in the form of well-rotted farmyard manure and other organics such as spent mushroom compost, but I will describe the methods by which I have had the greater success.

Cultivate or dig in the autumn and leave rough to expose the soil to weather and frost. In spring, and only when the soil can be pleasantly worked without stickiness and the implements and boots can be kept clean and dry, then break down to a fine tilth. This work can be done usually sometime in early March when drying winds and perhaps a little warming sun are about. There are always just two or three days which are ideal; nature has never let me down in this respect, it is just catching the opportunity. Liberal quantities of peat and perlite should be added, and gritty sand can also be used if the soil is heavy. The idea is to get texture and air into the beds. In my opinion this is one of the most important aspects of soil preparation and indeed in the production of those super spikes, and it can only be achieved if the soil is drying out at the time. Walking on the soil of the beds should be avoided; work from the paths. Raised beds are ideal, approximately a yard or so (1m) wide with paths between.

The pH can be tested at this time to determine the acidity or alkalinity, with a purchased kit or a small electronic meter. This pH should ideally be 6.5, slightly acid. If too acid adjust to correct with Dolomite limestone. This also provides calcium and magnesium, both minerals being very important to gladioli. The instructions in the kit or with the meter will inform you of the adjustment quantities. If slightly alkaline then peat should redress the balance, but if very alkaline then flowers of sulphur may be added. Just before planting or indeed at any time during the growing season further checks and adjustments can be made.

The next test should be for nutrients and I recommend a Sudbury Test Kit, which includes full instructions to determine the level and balance and any deficiencies of the nutrients in the soil. The subject of gladiolus nutrition and the application of fertiliser will be dealt with in detail later, but I stress once again that the most important aspect of soil preparation is texture and air, sponginess and lightness. The roots will thrive in these conditions, especially the fine feeding roots. Extensive root systems will not be developed if the soil is stagnant.

I find good quality peat and the clean light crumbs of perlite ideal, and the expense well worthwhile. It is no use buying expensive stock if the conditions are unsuitable. I have not used any manure or anything different from the above for the past 25 years. Rake the soil level and leave well alone until planting time.

## Obtaining the Corms

Of paramount importance is obtaining and planting good corms. When growing for maximum potential only the best selected corms of the best stock should be used. Gardeners refer to 'good strains', which are the result of selection and reselection of the propagated basic stocks, and should also be disease and virus free, as they say 'clean'.

In my opinion the very best way is to produce your own corms by propagation as described in Chapter 3. By this process you have the full process under your entire control and can practise full and intense selection, propagating only from those plants that have produced those very special spikes, and from those cultivars that have been found to be suitable for growing in your own particular environment and therefore have become fully acclimatised to the conditions. Another reason is that the corms are available for planting when you require them and you are not waiting for them to be transported from abroad and despatched from suppliers. Those imported from the USA and Canada can be a problem in this respect. To be given substitutes at the very last minute is also very annoying indeed.

However, it is not all that easy to produce your own corms in most parts of the British Isles, and only for the very enthusiastic and dedicated as a long-term programme is it required. But it is the only way if it is the intention solely to exhibit those cultivars which have been bred by oneself. Most of us are not in that fortunate position and have to purchase some new corms each year, and we always want to keep up to date by growing at least some of the best of the recent releases.

Obviously the best advice is always to go to a specialist supplier of gladioli, someone who is well aware of the requirements of the exhibitor, of the scene worldwide, of suitable cultivars, and of the best growers and suppliers of corms. The British supplier imports the corms usually from Holland, USA and Canada. Send for the catalogues very early indeed — they are usually available in September or October — and get your order in straightaway. I have received a catalogue in the morning post and had my order away that same day. Remember that the first orders get the first choices. Good corms in the requested cultivars are always difficult to obtain, so at least be at the front of the queue.

Some exhibitors actually send direct to the growers in America for their corms and there are advantages and disadvantages in doing so. It is however necessary to send a reasonable order to make it worthwhile, and there is the problem of dealing with plant health certificates, import duties and the transfer and exchange of money values. The certificates are sent by the grower, but if not all is perfectly correct in this respect then the corms will be refused entry into the country and destroyed. The money situation will be dealt with by your friendly bank as long as there is sufficient money deposited in the account! But be prepared to pay the postman the import duty when he presents the parcel on the doorstep some morning and the dog is snarling at his legs. Otherwise he will take it back to the post office.

The advantages are that you may save some money on a decent-sized order and that American growers usually like to impress their friends in England by sending good quality corms, and also may include quite a few bonus corms of the very newly released and perhaps indeed some generally unreleased cultivars.

Now what to look for in a corm. The biggest is not always the best. A young, plump, high-crowned corm is usually better than an old flat one (see Figure 8). Remember that vigour is usually on the side of youth. The age of the corm can actually be determined by the diameter of the root scar at the base, a small scar indicating a young corm probably propagated from a cormlet planted the previous season. In America the exhibitors usually stipulate to the grower that the corms are required for the purpose of exhibition and that young selected 'jumbos' are required. Of course they pay a few more dollars for the privilege. In the British Isles this is not possible unfortunately, but should not be so if there is a demand created and the buyer is prepared to dig deep into his/her pocket. Corms are expensive owing to transportation costs, import duties, handling and postal charges and the like, and specifying jumbos would probably put them in the 22 carat bracket and require hall-marking!

On receipt the corms should be clean and not show any signs of disease; peel them immediately on receipt to inspect. If any are sent that are obviously diseased then wrap them up again and send a 'many happy returns' parcel to the supplier. Remember, you have paid hard-earned money for those corms.

Most cultivars do perform best from reasonably young corms as described, but there are a very few exceptions when older corms are preferred. It is also a fact that it is necessary with respect to a few others, in order to grow really good spikes, that very young small corms are required, older ones being completely useless. The cultivars shown in the Pedigree Chart in Chapter 7 on breeding hybrids being an example. I have known top exhibition spikes to be produced from corms only $\frac{1}{2}$ - $\frac{3}{4}$in in diameter (12–20mm approx.) Really, it is just knowing the growing requirements of each cultivar through experience. Certain exhibitors have become synonymous with particular cultivars, always seeming to get better results than anyone else. For example, no one could grow better 'Landmark' than Ken Brough.

## Preparing and Conditioning the Corms before Planting

This is an important aspect that is frequently sadly neglected. Unfortunately, in many cases when the corms are received from the supplier they are examined once on arrival and then left in the paper bags and practically forgotten until planting time. I can assure you that more attention is required if you want to be in contention for that trophy at the big show. First of all peel completely each corm, being careful not to damage any shoots, and then examine the base scar. Determine from the condition of the root buds if the corm is still in a dormant state or that the roots have begun to swell to show that they are waking up from the winter hibernation. The corms should not be planted in a dormant state especially into cold, wet soil. The chances are that many of them will rot and fail to grow.

Therefore condition them before planting by placing in a warmer room for a short time. Use either seed trays or cardboard boxes or the like and place in a single layer and bottoms down. No peat or any other material please, just dry trays. The idea is to get the roots just showing but not growing, and the shoots to emerge upright and straight. It has been shown conclusively by scientific research that the plants will perform a lot better if the corms are stored upright. A little insecticide powder mixed with a fungicide powder will be an insurance against pests and any moulds. Greenfly and other insects have a habit of hatching and multiplying at an alarming rate in the warmth, and feeding on the young and tender shoots if this precaution is not taken. Hexyl is recommended as it is a ready-to-use mixture of insecticide and fungicide. Inspect frequently, and as we are growing for exhibition it is recommended that only one shoot per corm is retained. Remove the others very carefully with the tip of the knife, wiping between each operation and preferably on a cloth with sterilising spirit or Dettol solution. This stops the spread of viruses. Once planted they should grow away instantly if the practice of conditioning the corms beforehand is carried out.

## Planting

As described previously, it is essential to get the soil into a good and friable condition, with superb texture. The moisture content is also very important and this can be controlled if the land is heavy by covering the beds for two or three weeks before planting with clear polythene sheeting. This may be particularly beneficial in the northern parts of the British Isles. The edges can be held down with bricks. You will also now see another advantage of using the raised-bed system; the rain runs off the sheeting and onto the paths. The polythene also warms the soil and makes planting easier and a much more pleasant occupation. It also enables the planting to be carried out on any fine day by just rolling back the sheeting, and obviously ensures a good start to growth. The polythene can be left in place over the planted corms for a couple of weeks if desired. The golden rule is never to plant in cold wet soil.

It is recommended that the corms are soaked in a mixture of two fungicides before planting. I have used both Captan and Benlate for a number of years but as Captan may now be difficult to obtain another suitable fungicide can be substituted. Read the instructions on the pack for suitability for the purpose, the strength and soaking time. Plant immediately afterwards.

It is my opinion that most glads are planted too deep. Far better results are obtained in my experience by planting no more than 2 in (5 cm) deep. If the soil is heavy a handful of gritty sand under and over the corm may be beneficial, but I like to use perlite. Glad roots love to grow in the clean, white, warm grains, which must give the plant a flying start. The first roots are very brittle and delicate and must find a cosy environment on emergence. They must never come into contact with excessive or strong fertiliser; there is plenty of nutrient in the correct proportions stored in the corm.

As regards planting distances then this can vary according to the size classification of the cultivars being planted. I have described that the width of the raised beds should be about a yard (1 m). It is possible to grow three rows to a bed and I have done so very successfully in the past. I now prefer two rows but with the corms closer together in the rows. Two rows also make for more accessibility for staking, tying and other tasks later in the season. This way there is no more stretching through to the middle rows with possible damage to those precious spikes. Giant and large-flowered cultivars (500 and 400) need planting approximately 9 in apart (23 cm), medium-flowered cultivars (300) approximately 7-8 in apart (18-20 cm), small and miniature-flowered cultivars (200 and 100) approximately 5-6 in apart (13-15 cm). The best primulinus I ever grew were practically touching one another down the rows.

Make sure that there are no stones or lumps of heavy soil over the corms to impede the young, tender shoots. They require to be grown straight right from the very beginning. Do not carry out any hoeing or cultivation of the soil until all the shoots emerge so as to avoid damage, as once damaged they never recover. Slugs can be troublesome at this time and the use of slug pellets is a wise precaution. If you live in a rainy area alternate pellets with liquid Slugit (aluminium sulphate). This product remains active in rain and also kills the eggs.

## Timing the Blooms for Shows

Let me be honest and state right from the beginning that there is no magic method or technique available that will guarantee blooms for any particular show. Definite timing is impossible. Probably the most frequent question that I am ever asked regarding gladioli is for the date of planting to enable the blooms to be just right for a particular show. I can give a few suggestions that will increase your chances of having some blooms available. Making a list of the factors that influence the time of flowering is helpful.

1. The weather
2. The planting date
3. The type of cultivar
4. The age and size of corm
5. The preconditioning of the corms before planting
6. Depth of planting
7. The temperature just before and during the flowering period
8. Length of daylight
9. Luck!

To consider these factors in detail:

1. Obviously no one can forecast what the weather will be throughout the season. The weathermen often think they can make long-term forecasts and predict the summers, but with little success. Some springs start very dry and hot and growth is rapid, but by the time July comes around a far different pattern can and usually does

emerge, and long periods of cold and wet set in. The emerging spikes seem to hang about for weeks and seemingly give no promise of ever blooming at all. On the other hand, I have known spikes to be opening very rapidly indeed in the hot sunshine and practically the whole batch together, the blooms coming and going so fast that there seems no chance whatsoever of having any left for the show which is only a week away.

But this year the weather in our particular locality has been kind, and the main batch or 'cut', as an exhibitor refers to it, just coincides with the show date and we arrive at the venue with many wonderful spikes to the acclamation of our friends. This is *the* year, so enjoy the success. The biggest problem seems to be transporting all the blooms to the show and all those trophies back home!

2. The planting date is important but perhaps not as influential on the flowering time as first thought. A little time may be gained by planting early, and flowering can be delayed a little by planting late. But the results of an experiment that you may wish to repeat are worth noting. Obtain a number of identical corms of one cultivar. Plant them over a period of four weeks, and you will discover that the difference in flowering time is not four weeks but more likely days.

3. Some cultivars have a shorter natural growing period than others. A good catalogue informs as to whether they are early, medium or late flowering. I have observed over many years and by examining different cultivars that this can be related somewhat to the number of leaves that a particular cultivar grows before spike emergence. Some of the late flowerers have eleven leaves. As a very general rule the Dutch cultivars tend to flower before the Americans and Canadians. It depends to a great extent in what climatic conditions they have been bred and selected, and therefore acclimatised. The conditions in the British Isles are more like those in Holland than those in America.

4. Other things being equal, old corms flower earlier than young, small corms.

5. Corms should not be in a state of absolute dormancy on planting, but preconditioning is important. A corm with the root buds visible and the shoot beginning to grow will flower earlier than those in a less advanced stage. Advantage can be taken of this by 'chitting' as referred to in Chapter 7.

Another method sometimes employed by certain exhibitors attempting to get a particular and very late-flowering cultivar on the showbench is by planting early in peat pots in a greenhouse and then planting out in the beds later. It is important not to do this too early or a check to growth may result, especially if root disturbance occurs — that gladioli resent. Usually a small number of buds on the spike is sacrificed by planting in pots, but it could be beneficial just to try a small batch of that very special late-flowering cultivar that would perhaps otherwise be too late for the shows.

6. A deep-planted corm will take a few more days longer to flower than one more shallowly planted; again all other things being equal.

7. Once the spikes have been formed and just showing colour, the temperature and atmosphere at this particular time influences when full inflorescence takes place. If the weather is cold the spikes seem to remain in the same stages for ages and struggle to even open, sometimes only one floret at a time, and taking nearly as many days to open the approximate required eight. If the temperature is hot and the atmosphere bouyant then three or four at a time open, taking just two or three days.

8. The difference in daylight hours does vary from the south of England to the north of Scotland as does the intensity of light. This, allied to the difference in temperature, does effect the flowering time. There is no doubt that the climatic conditions of some locations do favour the requirements of gladioli and produce good spikes that have 'weight' and superb texture of floret. Certain areas have become associated with a number of successful exhibitors who have consistently produced top exhibition gladioli over the years.

Who can ever forget the exhibits of Ken Brough, Joe Hewer, the Waltons and other top exhibitors from the Carlisle area? They were all very skillful growers and I give all due credit to that, but I am convinced that the microclimate of the location had a certain influence in producing those wonderful spikes.

9. And we all need a bit of luck!

From all this perhaps we can evolve some guide lines to give the best chance of having at least some spikes ready to cover that particular show or better still a number of shows throughout the season.

(i)    Grow a reasonable number of corms of a reasonable number of different cultivars and include early, medium and late flowering, some of Dutch origin and some of American and Canadian and, if you can, get hold of some of British origin.
(ii)   Try to have two different plots of ground and if possible at different locations.
(iii)  Stagger the planting times as much as reasonably possible.
(iv)   Start a few corms under clear polythene sheeting on ground that has been pre-warmed for 2 or 3 weeks before hand.
(v)    Pre-chit a few corms.
(vi)   Keep record of planting times, flowering times and climatic conditions for your particular locality. These will be invaluable in future years.
(vii)  I hesitate to give actual dates to commence planting as I obviously have not the experience of growing in all parts of the British Isles, but in my own locality in Lancashire the second week in April is about right.
(viii) Get the prayer mat out at regular intervals!

**Gladiolus Nutrition**

This is a very complex subject made simple because all the research and experimentation has already been done. I immediately acknowledge the advice given to me by John Evans of Penrhyndeudraeth, Gwynedd, North Wales. John is a very knowledgeable and successful exhibitor of gladioli. As friends we have exchanged ideas, information and stock over a number of years. Being also a research chemist he is very interested in the nutritional aspects of gladiolus cultivation. The information given in this section is alone well worth the price of the book to the keen exhibitor of glads. Much of it is exclusive and not been released or divulged before. The best exhibition spikes I have ever grown have been fertilised by the methods described.

To ensure maximum availability and release of the fertilisers that are to be used it is essential that the pH of the soil is correct. This should always be checked before planting and adjusted by the application of Dolomite limestone to 6.5. It is also necessary at this time to check the nutrient balance for the main elements: nitrogen, phosphate and potash. The ideal fertiliser for gladioli expressed as a percentage and of the main elements is 8 nitrogen, 32 phosphate, 16 potassium and therefore in the ratio of 1:4:2. The composition of these elements should ideally be varied to take into consideration the existing balance of the soil, but to give the final 1:4:2 availability. The content of the fertiliser can be adjusted to accommodate the requirement.

The John Evans Formula consists of di-ammonium phosphate (20% nitrogen, 53% phosphate), nitrate of potash (13% nitrogen, 46% potash), *triple* superphosphate (44% phosphate). If these are mixed in the proportion of 1 part di-ammonium phosphate, 2 parts nitrate of potash, and 3 parts triple superphosphate we arrive at the following table:

| | |
|---|---|
| 1 part di-ammonium phosphate (20:53:00) | 20 : 53 : 00 |
| 2 parts nitrate of potash (13:00:46) | 26 : 00 : 92 |
| 3 parts triple superphosphate (00:44:00) | 00 : 132 : 00 |
| | 46 : 185 : 92 |
| Ratio | 1 : 4 : 2 |

The ideal ratio for gladioli! You can vary the composition of the elements to take into account the existing balance of the soil by making adjustments to the table.

The other elements that are important to gladioli are calcium, magnesium, and boron, with other trace elements in much smaller quantities. The calcium to some extent is supplied by the application of Dolomite limestone, used if necessary to correct the pH of the soil. Your particular soil may already be very rich in calcium but in some cases the calcium may need supplementing and this can be done late in the season just before flowering time by one application of calcium nitrate (nitro-chalk). Glads need calcium at flowering time. Magnesium is also present in Dolomite limestone. The boron and other trace elements can be supplied in the form of

Trace Element Frit 253A with the following analysis: 12% iron, 5% manganese, 4% zinc, 2% boron, 2% copper, 0.13% molybdenum, balance to 100% sodium and silica. All these fertilisers can be obtained from Chempak Products, Geddings Road, Hoddesdon, Herts, EN11 0LR, or from other reputable fertiliser suppliers.

All this seems rather complicated in theory but it is not in practical application. The fertilisation programme is as follows.

1. Adjust the pH before planting by the addition of Dolomite limestone if the soil is too acid.
2. Adjust the balance of the nutrients in the soil before planting by the addition of the modified John Evans Formula, use 2 ozs to the square yard ($60 g/m^2$).
3. Apply $1/4$ oz per square yard ($8 g/m^2$) of Trace Element Frit 253A before planting.
4. Apply 2 oz per square yard ($60 g/m^2$) of the John Evans Formula at the following times:

(i) four weeks after planting
(ii) the middle of June
(iii) the middle of July

5. Apply 2 oz per square yard ($60 g/m^2$) of calcium nitrate at the end of July but only if required.

If the weather is dry during the growing season it may be necessary to water-in the fertiliser applications to make them available to the plants.

One of the reasons for the effectiveness of the John Evans formula is that most of the elements are in a readily available form to the plants, and they do not require breaking down in the soil and converting as is the case with many other fertilisers, especially those of an organic nature. They do not rely on the temperature of the soil and this is important in a cold spring. Also flower quality is better when nitrogen is supplied in two forms, nitrate and ammonium.

A gladiolus corm, unlike most bulbs, has not the embryo flower spike formed before planting. It is only formed some 5 to 7 weeks after planting, and it is at this point that it makes its mind up how many buds it will have on the spike. This is very important indeed to the exhibitor. If nutrients are not in a readily available form at this critical period then the result is lack of buds. I have never been short of buds in any season using the John Evans Formula.

The base fertilisers used before planting should be well worked and distributed within the soil, as initial roots are very delicate and concentrated fertiliser in the root zone will burn them. The nutritional programme has been devised principally to produce those top-quality exhibition spikes in the giant and large-flowered classifications. The amounts of fertiliser should be modified when growing small-flowered and miniatures, which should not be overgrown and forced out of character. Primulinus require even less. It must be emphasised that overfeeding should not be carried out whatsoever. It is a great temptation to apply just a little bit more, but I can assure you that a full season will be wasted. I have carried out many experiments over the years to determine the amount and content of fertilisers and have purposely overfed just to discover the effect. Believe me, the results have been disastrous.

## Seasonal Care and Attention

Once planting has taken place there is little to do until the shoots begin to appear and break the soil surface. Cultivation up to and at this point should only consist of any necessary hand weeding to keep the beds clean. No hoeing or even hand forking must be carried out for fear of damaging those precious shoots. Once damaged they never recover and may even rot. This also applies to slug damage, so keep the slug-killing agents handy and in use. Young gladioli shoots are a very attractive and tender meal to the slug or snail.

Cultivaton of the soil surface can be done when all the plants are growing away. Hoeing is very beneficial but make sure it is only the very minimum depth so as not to cause damage to any of the fine roots. You will also appreciate having just two rows per bed, making cultivation a lot easier than having three rows. And the bed method avoids walking near the plants, resulting in compaction of the soil. Regular spraying with fungicide and insecticide is very necessary.

A decision has now to be made as to whether a mulch should be used. Definitely not one of manure however well rotted, and certainly not one of grass cuttings. The choice is endless of other materials. I have used peat very successfully on many occasions but the material that I consider the very best is pine needles; yes, pine needles! Research has shown them to be very desirable indeed for a number of reasons. Their use has not really been exploited to the full in horticulture as yet and they don't seem to be available unless you collect them yourself. The needles are taken from the floor of a pine forest and are usually of a very thick carpet. They have definite disinfectant qualities and help to prevent disease and keep down the eelworm and nematode count. The soil surface is protected from panning by heavy rains and weeds are easily removed by hand, if in fact they grow at all. There are usually no weeds and weed seeds on the floor of a dense pine forest and so are not imported in the material. Watering and feeding is very easy under these circum-stances and unlike peat which once dry is difficult to rewet. The mulch breaks down into excellent humus when dug into the soil in the autumn, providing a long-lasting texture.

A few pleasant car trips into the countryside can be combined with collection but I do ask you not to just jump over the wall and into the first pine forest you see — you wouldn't do that would you? There are lots of pine forests around reservoirs, and contacting the friendly bailiff first will usually do the trick and get the necessary permission. Usually there will be no problem collecting a few sacks at a time. The material is very clean and will not dirty the boot of your car. All in all pine needles make a superb mulching material and are well worth going to a bit of trouble to obtain.

Another excellent but unusual mulching material is bracken peat, which can be used in a similar way to pine needles. This is the spongy peat from bracken woods. The only problems seem to be the acidity, but this can be corrected by application of Dolomite limestone. I have used bracken peat extensively as there are plenty of bracken slopes very near to my home and within wheelbarrow distance.

It is usual at some time during the summer to experience a few heavy winds and these may push the plants over from their straight and upright position, especially if planted shallow as I recommend. Normally they can be eased back upright again and the soil firmed around the base. They must grow straight or otherwise the outcome will be twisted spikes which are useless for exhibition. In very exposed situations it may be necessary to insert a short cane about 2 ft (60 cm) long and away from the plant but at an angle across it. Soft string can be looped around to support against heavy wind – a wise precaution. It may seem a lot of trouble but is soon forgotten when carrying off that trophy! The main staking of the spike will be carried out later and should not be attempted now.

If the weather is really dry during the summer and your soil is light or gritty, watering may be necessary but do not overdo this if drainage is not very good. If the water vanishes and drains through easily then overwatering cannot really be done, but otherwise be careful or a quagmire may result and destroy all the previous good work in creating texture, tilth and air-pore space in the soil.

June and early July are ideal months for the gladiolus exhibitor to take a holiday before the show season starts, and he or she can go away knowing that the precious plants will take care of themselves for a few days. With international travel being much easier these days why not take the trip of a lifetime to the United States of America? There are some super gladiolus shows there! The publications of the North American Gladiolus Council will give all the details.

## Development of the Bloom and Staking

The exciting time arrives when the base of the plant which has up to now been just a flat sheath of leaves begins to thicken and round. The spike is on its way! I have never ceased to be amazed at the rate of growth of the gladiolus as the inflorescence develops to full bloom. From a mere 2 ft (60 cm) or so tall, the spike reaches up to some 6 ft (1.8 m) or more in height in the giant and large-flowered classifications and only in a very few short weeks – some growth!

What the plant needs at this stage and to support all this growth is moisture. It does not require any supplementary feeding whatsoever as the diet has been supplied in a balanced and controlled form as detailed in the chapter. Do not upset the good work as any last-minute, so-called tonic will do more harm than good. It can cause uneven placement of the florets on the spike, floppy and poor texture petals, and can even make the tips of the buds like rubber. You have been warned: there are far too many exhibits ruined at this time and a whole season's work wasted by unnecessary feeding.

Staking is necessary, but not before the buds emerge from the sheath of leaves and show clearly the way the spike will eventually face. Watch the very tip for the indication as it points forward. The cane is inserted behind and upright and taking care not to damage the corm. Occasionally we do get it wrong and it is eventually shown that the cane is at the front and has to be replaced — not to worry. Release the flower-head bud if it gets caught up in the leaves.

I must announce here and now that exhibition gladioli do not develop by themselves into those tall, majestic spikes seen at the specialist shows. Whether it is considered unnatural or not, if your ambition is to be successful you need to 'dress' the blooms. Dressing commences at the very time the cane is placed into position, and as it stretches upwards it must be fastened by use of twist-it ties. A tie at every other space between buds is not too many. Crooked spikes are useless for exhibition, but are better given away to the floral artists where they will be appreciated more. Now is the really busy time and daily and constant attention is required. In warm weather development is so fast that it is essential that the ties do not become snagged resulting in 'shepherds crooks'. It is also at this point in the season that you may suddenly discover that too many corms have been planted.

## Dressing

Eventually the whole of the spike will have stretched and be tied to the cane. Now is the time to ensure that the buds are in position to allow the florets to open all facing forward, and correctly spaced to form as near a perfect spike as possible and in accordance with the requirements for exhibition. If the spikes are not dressed and the attention given then, it is very unlikely that you will achieve any success in a good standard of competition.

The buds can be manipulated to achieve this, but experience is a definite advantage and can only really be gained by practice. Florets should never be manipulated when they are turgid or they can easily be damaged or may even be snapped off. Early morning, late evening or in cold weather are the times to avoid. Midday to early evening is the best time, when the buds are soft and pliable and it is possible to move them about and into place more easily. The exercise should be repeated daily on those blooms that have been selected as candidates for a particular show and continued to the time they are cut, when cut, and also when staging and preparing the exhibits actually at the show venue. Florets can actually be opened very quickly from a partially open to a fully opened bloom, and the spike balanced to give the required number open, showing colour and unopened buds..

The best advice I can give is to take any opportunity to watch the regular and successful exhibitors prepare and stage their exhibits at a specialist show, usually overnight or in the early hours of the morning. This is an experience and well worth missing 'a few hours sleep for. I don't know an exhibitor who minds being quietly watched. The other alternative is to ask for some advice at an information table at one of the shows administered by the British Gladiolus Society, when some of the techniques can actually be demonstrated for you on a spare spike or two. Usually one of the successful exhibitors who is about will be willing to do so.

# Cutting the Blooms

At the countdown to show day you will find yourself very busy indeed as all the spikes commence to open their florets and require daily and constant attention. Tying and dressing becomes an enormous task but should be pleasurable in anticipation, especially if there is some good material to work with: long, healthy and many-budded spikes. One finds oneself listening to every weather forecast and anxiously scanning the skies to try to determine weather conditions. I can assure you that conditions will never be just right and perfect!

If the show is say on a Saturday then by the previous Monday and Tuesday we should have some idea which spikes could be available and in contention. From then onwards these are the ones that should receive the best attention. When to cut may be a problem, but should really not be so. Do not be afraid to cut. Fortunately gladioli will open well in water after a certain stage in their development has been reached. This can vary with the different cultivars as some will actually open from tight-bud stage, but others may struggle. Once the bottom floret has fully or very nearly fully opened then all cultivars can safely be cut, when the spike will continue to open. If the weather is wet and windy at this stage or on the other extreme very hot, then I would cut from Wednesday onwards. The advantage of this method is that the blooms can be kept clean and protected from the weather when opened indoors. Again, experience is invaluable in this respect.

As soon as cut from the plant, place the spikes in water, even if only temporarily. Once the batch has been cut and safely taken indoors further work can proceed. Where to keep cut spikes is a question often asked. Everyone can find some suitable cool and not so bright a room or building. I keep mine in a well-ventilated shed in the shadow of trees on which the sun does not shine. It is cool and not too bright although there is some light coming in from the front. Cellars are usually good if the air circulates sufficiently, as is a north-facing room in the house. A friend of mine keeps his in an old air raid shelter partially underground but covered by earth. Another gladiolus exhibitor I know keeps his spikes in a dairy and another in a cave in the hillside! I did hear a remarkable but funny story of someone using an ice-rink but the evidence showed by the skate marks on the lower petals! Of course if the weather is cold and the show date fast approaching it may be necessary to bring some of the spikes into a warmer and lighter place in which to accelerate development.

The 'walk-in' cooler is used a great deal in America when the weather is very hot, and spikes can be stored in one for quite a period of time to obvious advantage. Although not normally necessary in the British Isles, it has been known for one or two exhibitors to obtain the use of a refrigerated room from a friendly shopkeeper, supermarket or butcher, but the temperature and humidity must be correct for flowers or otherwise there are more problems than advantages.

When cut and placed in the chosen location for opening it is essential that the spikes are stood in water, upright and directly facing any light source, however dim. Unless this is done then crooking will occur and the florets will have difficulty facing forward. The best way is to fasten back again to canes by twist-it ties and on these they will remain during transportation and up to actual staging time at the show venue. The addition of some floral preservative may be a help in the water, the same as sold by the florist to keep cut flowers fresh longer. There are quite a few proprietary makes available. And remember to continue to dress those spikes as the florets open to ensure correct placement and facing.

## Artificial Protection

Would it not be better to leave the spikes to develop naturally on the plant until just before the day of show, but giving some form of overhead protection? Yes, it is possible and there are a few methods available and well tried. They all demand much work and attention but could be well worthwhile, especially if you are keen and enthusiastic. I have won at the highest level with protected spikes but I have done equally as well with the pre-cut tactics.

Many visitors to a large gladiolus show exclaim that the blooms are so clean that they must have been protected somehow. The truth is that some have but most have not! Certainly they are really useless for exhibition if grown in a greenhouse as they will become weak, drawn and have poor floret texture, the bud count being also usually very low. Experiments have been taking place recently by growing in the polythene tunnels so favoured by commercial growers. The most suitable designs are those with fine mesh sides at the base and ends to allow maximum ventilation and to avoid condensation. In fact it is best to leave the ends·open during the growing period. The biggest span and therefore the highest headroom types are best of course. Some of the small-flowered and primulinus glads do quite well under these conditions, but the ultimate would be to grow within an uncovered tunnel structure and just to put the polythene protection in place for the very last few days when the first buds begin to show the first colour.

The dahlia and early-flowering chrysanthemum exhibitors grow in outside beds but under framed structures, and then cover with top lights or polythene just at the flowering period. I have actually done this myself with gladioli some years ago with tremendous success. But exhibition, large-flowered glads grow tall and an enormous structure is required, so perhaps it is better to stick to the miniatures! My own covers were 7 ft 6 in (2.25 m) high to the eaves and 8 ft 6 in (2.6 m) to the ridges! My present tunnel is 9 ft 6 in high (2.85 m) to the ridge.

Attempts have been made to utilise polythene bags stretched over canes to protect individual spikes. This is a great idea in theory, but then the sun comes out when you are away at work and conditions inside that sweat box become intolerable and the blooms are ruined. A well-respected old grower and lifelong friend used to cover his cane with wigwam frames with cheese cloth or 'fents' obtained from the cotton mill. Now these were good — his plot looked like a Chinese kite festival but his haul of trophies at the weekend proved the effectiveness of the method.

Many of the old exhibitors constructed traditional 'bloom boxes' made of plywood on three sides and top, with an adjustable or removable glass front and shading. The box was designed for individual blooms and was fastened securely to a robust wooden stake. A lot of trouble perhaps, but at least one currently successful exhibitor admits to using them. Using them on selected spikes means that only a few are required, perhaps ten will do the job; after all, only one spike is necessary to win the Grand Champion at a particular show!

A very simple but very effective method of keeping those lower florets fresh and clean is by protecting them with a tube made out of ordinary brown paper. The tube should only be of sufficient diameter to just cover the florets and be easily fastened to the supporting cane. Personally I have not used brown paper but have used very successfully indeed a tube made out of aluminium kitchen foil. This lasts better than paper and can actually be squeezed around the supporting cane and clipped with a fulcrum clothes peg. Aluminium reflects the sun and keeps the florets cool. The top of the tube can also be shaped to fit the opened buds and stop the rain entering. The tube method can also be very effective in stretching spikes. It is so effective that care must be taken in its use so as not to do too much and upset the natural balance.

## Transporting to Shows

Over the many years that I have been exhibiting I have used all kinds of transportation and tried all different methods of packing. Much care and attention was lavished on those precious spikes. Some journeys were very short and of only a few miles, but in contrast some journeys were practically the length or breadth of the British Isles. The conditions encountered were also in the extreme, at times very cold and miserable, at times very wet and humid and particularly on one occasion I travelled some 300 miles (483 km) during one of the hottest days on record in England.

I have seen all kinds of weird and wonderful contraptions, boxes, racks, structures and methods of travel. Of course there is always the 'I may not have the best spikes on show but my bloom boxes are something else' brigade. The Americans go into unbelievable arrangements to travel to shows, and plan the campaign like a military exercise. I read an article by a well-known exhibitor who actually changed the shock absorbers on his station wagon just to ensure a softer and smoother ride. One thing that I have discovered is that gladioli are very tough and will endure travelling very well indeed. If reasonable care is taken then few spikes will actually be damaged.

The best way is to place them flat, florets upright, with the spikes still fastened to the canes at the time of cutting. They can be packed close together, in fact this may be an advantage as they support one another, but they must be prevented from sliding about. This can be achieved to a great extent by not having a smooth and slippery surface on which to place them. An old blanket may be useful in this respect. If many spikes are involved and carried, then boxes made of hardwood, plywood or cardboard can be employed and stacked in the vehicle.

When travelling long distances in hot weather it may be advisable to keep the cut ends of the stems moist by some means, to prevent dehydration. I usually use Oasis cut into small cubes and soaked in water and stuck on the stems. A small square of polythene sheet and a rubber band or some cling film completes the job. For shorter distances it does not matter; in fact some exhibitors actually prefer to allow the florets to just soften by keeping them dry. They then find that it is easier to dress the spikes when staging. The florets are also easier to open when in this condition. Some experience is obviously necessary if this technique is employed, but I can assure you that it can be very effective. The blooms soon recover their crispness when staged in the vases and given a drink of water, and of course they retain the formation and placement to which they have been dressed. However, in the end the best method is the one that fits your own particular purpose and the facilities which you have available. I repeat that gladiolus blooms are surprisingly very tough and durable.

But what we all hope for when leaving home to travel to a show venue is some really good basic material with which to work, and then we have a chance. Actually, the better-grown material can withstand travel and handling better as they will be full of vigour and have wonderful floret texture.

## On Shows and Showing

The apprehension of the newcomer to exhibiting is understandable, but I encourage him or her to have a go even at a specialist gladiolus exhibition. There are novice classes which can be attempted at first, but many newcomers have entered for the very first time and come away with a top award in a top class. I can quote many examples.

The main thing is to take advantage of the opportunity and to gather all the information available — and there is nothing quite like actual experience. Help and advice is always on hand as gladiolus exhibitors are very friendly people, and I don't know any of the current successful exhibitors who would refuse to give advice to a newcomer. Don't assume that your spikes will not win and you have no chance. I'll let you into a secret; most successful exhibitors leave home with apprehension and thinking their spikes are not all that good.

On entering the showroom or marquee and seeing the rows of entries do not let fear overcome you; the temptation to beat a hasty retreat back to your car should be avoided. They all look good at the first impression when staged; believe me, once yours are dressed, staged and presented properly they will look equally as good as most. Over the years I admit to having won quite a few cheap prizes by studying the entries and opposition, and exhibiting where I know I have a chance to win with what I have brought.

I have some exciting memories from my early days exhibiting at Southport. The last few minutes just before the stewards cleared the tent for judging was mad panic. Exhibitors would swop and change their entries around and around again, just like the runners in the last lap of the 5,000-metres race.

And here may I make a plea to organisers of gladiolus shows. Why, why do you require pre-entry many days or even weeks before the event? Please allow entry to be made during staging time. I know you require pre-entry to enable you to allocate space on the benching to each class and in accordance with the entries. I know you require pre-entry to make out all those neat entry tickets and envelopes and secret exhibitors numbers. But do you know that you are actually turning potential exhibits away? Many is the time I have pre-entered certain classes believing I have the material to exhibit. The weather has then changed and all the plans are altered. Those six spikes I had planned to enter in the big class are not all open due to a cold day or two, but I have managed three. But I have not entered the three-spike class. Where can I use these three spikes? Well I suppose I can split them and exhibit in the single spike class which I have luckily pre-entered. But only one entry per class! So two good spikes are wasted. What a disaster!

Not to be outdone next year I pre-enter every class in the schedule. It costs me money but at least I know I can somehow use all the material I have available on the day. The show secretary is very pleased as he gets hundreds of entries from all those with the same ideas. He spends night after night making out those neat tickets and envelopes, most of which end up as rubbish on the floor. He has also wasted a lot of time setting out his show benches to accommodate the pre-entries in the separate classes, some of which require huge spaces — and somehow they never materialise.

Gladioli are not easy to exhibit and many factors are against us, so do please try to make it a little easier. We only normally get one flower spike from each corm each season; they are not 'cut and come again' flowers. There is a lot of wasted material between shows. We do want every spike we can possibly get actually on those show benches. Keep the schedule simple and to cover at least 3 spikes, 2 spikes and 1 spike in all categories and classifications, please. And how about a footnote at the end of each schedule of the gladiolus show. 'Please bring your spike along, if there is not a class then we'll find you one!'?

And please, let us have less fuss with entry cards. All it requires is a piece of card with the exhibitor writing his name on one side placed downwards tucked under the vase and the class number upwards — and the exhibitor does it all for you when staging! The judge will not look at the name until all the awards, including the major awards, are judged. It won't matter anyhow as it's the glads that get the awards, not the exhibitors.

I could give you all the usual advice about reading carefully the schedule regarding the classes, staging times, judging time. Are the vases provided or do you have to supply your own? All very boring stuff but very necessary.

## Judging

Judging gladioli is a fascinating and absorbing interest in itself and very rewarding. To be actually invited to perform the function at a large or specialist show is indeed an honour. Judges are in the fortunate and privileged position of actually being able to examine every spike in every detail and therefore to be very familiar with all the cultivars and all the new introductions.

Unfortunately there are not many really good judges about. This is a serious business and a lot of responsibility is placed on a judge. The exhibits should be given the precise attention and considerations they deserve. Much care and attention has been lavished on them throughout the growing season, travel, preparation and staging at the show, not to mention the costs involved. It would be fatal even to contemplate just what each spike has really cost. Yes, the exhibitor is pinning all his hopes and aspirations on the judges' decision and does and should expect to have fair and careful consideration.

A good judge is impartial to particular colours, types or forms or even certain favourite cultivars, but should always be on the look out for new and improved developments. Unfortunately, many judges are 'knockers' when they should really be 'encouragers'. Their method of approach in judging is to place the exhibits in descending order of faults instead of ascending order of excellence, demonstrating the virtues and qualities, excellence of cultivation, preparation and staging. What an attitude to go with into a show venue to judge, looking for all the faults; what a miserable and unenjoyable task! It is a far greater pleasure to look for the better features and seek out the best spikes and those that have obvious quality and are to be admired. In my opinion, the best way, if you have aspirations to become a judge, is to carry out the job of a steward for a few times at a gladiolus show. This can be very valuable experience, especially if stewarding for a number of different judges. Usually by approaching the show secretary some time in advance and by showing an interest, arrangements will easily be made.

Eventually some tuition could possibly be arranged with a qualified and experienced judge to prepare for entrance to the British Gladiolus Society examinations. Even so, there is usually a training session held immediately before the examination and on the same day. But it is best to regard this just as revision; all the studying should have been carried out previously. The great majority of those wishing to become judges attempt to pass the examinations before being actually prepared. What other examination would an entrant sit without proper preparation and study? Perhaps the would-be judge has been a prominent and successful exhibitor for a number of years and assumes that the experience he has gained will surely allow a pass mark. In most cases this is not so. I must admit that I fell into this category myself and failed first time, and it really took some soul-searching and a complete change in attitude to pass the examination, let alone become a confident judge.

Remember to be constantly making reference to and studying the *Points System for Judging* booklet published by the British Gladiolus Society until you know every detail really well; but it is permissible to have the reference available when taking the examination. An enquiry to the Secretary of the British Gladiolus Society expressing your interest will be necessary, and full details of dates, venues and entrance details will be provided.

If you go into a show, having been engaged as a judge and,

seeing all those rows of magnificent gladioli in perspective and to a vanishing point, suddenly have the butterflies and want to beat a hasty retreat, then perhaps it would be better if you did so. You must have complete confidence in your ability, and if you therefore do decide to stay then get to grips with the situation and proceed in an orderly, methodical manner and approach. Take your time, never be harassed or pushed. If necessary leave that particular class that you may be undecided about, as the competition is so very, very close. Carry on judging other classes and then return. Amazingly, you will usually find a decision is clearer and much easier than at first thought. Do not be too proud to carry out this practice.

It is not necessary for every spike to be 'pointed', in fact this is certainly not practical in the normal time allowed for judging. An experienced and competent judge can place the exhibits in order without, but in close competition and in certain classes it is a requirement. All judges should, at every show, apply the 'pointing' procedure to at least a class or two just to keep him or herself familiar with the system. Do not be too proud to refer to your booklet. In fact it should be carried with you for constant reference during judging. And of course it is necessary to have all official references and lists of 'Gladiolus cultivars classified for show purposes' available.

It is interesting occasionally to select a particular spike and elect that it is, say, an '84 pointer', or whatever. Then actually point up and see if you have forecast correctly or how far out you are. After some practice you should be able to determine how many points a spike would be awarded even before they are added up. This is a very good training exercise and also would check your consistency. Although never been done before to my knowledge in show reports, it would be an incredible advantage to actually record the points *against* a particular spike of merit. Reading show reports can tell us that the champion spike was a wonderful example of a particular cultivar. But how good was it really? If it could be quoted that it was awarded 98 points, then that really means something to a person reading the report who was not actually able to visit the show.

Do not be afraid to discuss your decisions after judging, in fact this to me is very enjoyable and a time to relax after the job is done. But never give any impression or indication whatsoever of doubt or changing your mind; this can be fatal. But do listen to other people's opinions; we can always learn a little more and sometimes from the most unlikely sources.

I must admit that I really enjoy judging the seedling classes. Being a breeder myself I am always looking for that very special seedling that could be extremely unusual and perhaps the forerunner of a different type or feature in the gladiolus world. It is very important to be extremely familiar indeed as to what is already available and has been achieved. In fact this is necessary so as to be quite sure that no existing classified cultivars are entered and are ineligible. The seedling classes should not be judged on the same lines as the other competitive classes. It is not just that actual spike as presented before you that requires judging but its potential, novelty or as a development or improvement on existing cultivars, and 50 per cent of the points should be considered for award against these factors.

Potential is not easy to assess. In my opinion only a breeder who has selected seedlings for a number of years can really understand and be able to predict potential — even then not always accurately! Would you retain that particular seedling or would you destroy it? is a question to ask yourself. And remember that seedlings should not solely be assessed for exhibition potential but for all kinds of purposes and uses. And don't forget your tape measure or ruler when judging the seedling classes. Unclassified cultivars are to be measured by floret size to place them in the correct and appropriate classes!

## References

*List of Gladiolus Cultivars, Classified for Show Purposes, with Points System for Judging*, published by the British Gladiolus Society

*The Gladiolus — Its Cultivation, Exhibition and Propagation, including Classification and Official Judging Points for Gladioli and Instructions to Judges*, published by the British Gladiolus Society

*A Selected list of Gladiolus Varieties Classified for Show Purposes*, published by The North American Gladiolus Council

*Dressing of Gladioli for Show Purposes*, published by the British Gladiolus Society

*Points System for Judging and Notes on Judging System. Non-Primulinus, Primulinus, Baskets, Floret Boxes*, published by the British Gladiolus Society

# Chapter 7    Breeding Hybrids

One of the most fascinating, interesting and totally absorbing aspects of gladiolus culture is the breeding of new cultivars. New cultivars are, of course, raised only from seed. The exception is by mutation or 'sporting'.

Imagine having a cultivar of your very own, knowing that there is not another quite like it in existence. The thrill of seeing those seedlings open for the very first time is indescribable, and you cannot wait for the florets to show a glimpse of colour. Have our dreams and goals been achieved? Perfection is not possible and we are always trying to make improvements by each successive generation. There are many disappointments along the way but the few successes are ample reward for all the efforts. Creation is a wonderful thing, and to have played some small part in that experience is satisfaction indeed! There is no doubt that some people have a particular 'gift' or aptitude for this kind of work, similar to a musician or artist. Everyone has some prospective talent, and it is the discovery and development of that talent that may lead to great achievements. Perhaps we may be allowed to help, in a small way, to discover and develop any talent you may have for gladiolus breeding.

You may have been discouraged by the length of time you believed it takes from making a 'cross' to actual flowering. Really, we do not have to wait too long, certainly no more than for most perennial plants, and if seed is taken and planted each and every year then new creations can be enjoyed each season. Maybe you have been put off breeding by reports of the vast quantity of seedlings that must be raised, and most discarded, to discover a very small percentage of worthwhile cultivars. This certainly is not true if a definite and well-planned breeding programme is formulated before commencement. I will endeavour to guide you through the establishment of such a programme in very practical terms. On the other hand, if you want just a bit of fun then there is nothing wrong with having a dabble and cross pollinating two glads that happen to be flowering at the same time. Don't expect too much, but you may indeed be lucky. It is possible to win the football pool jackpot with a one-line entry!

It is true that many breeders do raise vast numbers of seedlings and hope by the 'theory of quantitative reproduction' — the more grown the more chances — that at least some will be worthwhile. That in our opinion is not a systemised and scientific approach to breeding new cultivars. There are reports of planting seeds by the bushel and vast acreages of seedlings. Think of the costs involved. One breeder has a special seat attached to the back of a tractor and is driven around the seedling fields making his selections — that is

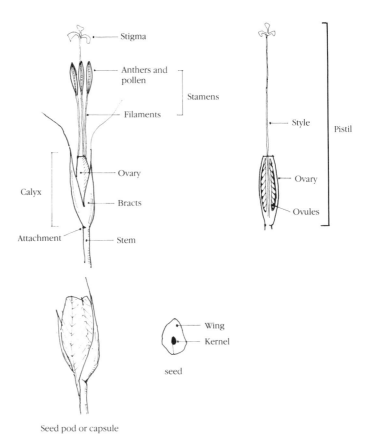

*Figure 10a Parts of the flower and seed pod*

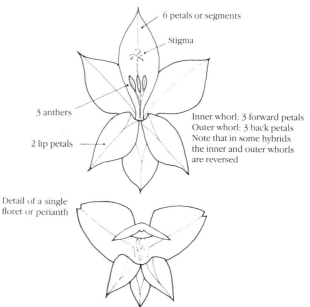

*Figure 10b Typical 'prim' hybrid showing hood and wings with blotches to show insect guides to nectary*

not really clever. It is far better to concentrate on the genetics instead of the mechanics! One thing is certain, the commercial prospects of making profit out of gladiolus breeding is very remote indeed. You will never become rich in monetary terms, but far more rich in other aspects of life.

## Basic Botany, Fertilisation and Seed Formation

Producing the seeds of gladioli is fairly easy, but it is necessary to study basic botany, fertilisation and seed formation. Parts of the flower including the reproductive organs are shown in Figure 10.

Each individual floret, consisting of six petals, possesses three stamens and one pistil. Each stamen terminates with an anther on which the pollen forms. The pistil terminates with a three-part stigma. When the pollen of one gladiolus is deposited on the stigma of another gladiolus, cross-fertilisation usually takes place; but if its own pollen is deposited on its own stigma and fertilisation takes place, then this is said to be 'self-fertilisation'. On fertilisation seeds are then formed in the ovary. The seed pod develops and eventually ripens and splits to reveal the seeds.

In the wild, fertilisation takes place naturally within a species, and the pollen is spread usually by insects or by the wind; but we must control this process within our breeding programme and maintain the records of the parentages. This is no different from the techniques of breeders of other plants and animals. For example, racehorse breeders carefully maintain the ancestry details by registration and stud book records. We can do the same with gladioli.

It is not always easy in all parts of the British Isles, especially in the north, to fertilise and set seeds in the open garden, and even so the shortness of the season does not allow the seed pods to ripen and mature properly. It is recommended that the work be carried out in a greenhouse. Many breeders are now using the modern plastic-covered tunnels for their work and these are very satisfactory indeed. The essential ventilation is provided by the facility to open the ends, and it is now possible to have partially netted sides. The growing environment is excellent. The area of protection afforded is most economical.

The programme should be preplanned as it is necessary to select in advance the cultivars one wishes to use that particular season. These are the breeding-stock cultivars, the selection of which is obviously very important indeed to a successful programme and result. The selected corms can either be planted in the greenhouse beds, in large pots, or by using the ring culture method. The idea is to get them to flower as early in the season as possible to allow a long period after fertilisation for the seed pods to develop and ripen in the warmer and lighter conditions, hopefully helped with a bit of sun. It is a fact that large, old, corms generally flower earlier than young ones. To take advantage of this, use your old corms that are past their best for normal flower production. Take them out of store very early and chit them in dry trays as you would a potato tuber; no plunging in peat or similar materials, please. A window sill in a warm room is ideal. Get the sprouts to grow but not roots — this is quite possible if the corm is kept dry. When the danger of frost has passed, some heat in the greenhouse is an advantage but not essential, then plant the corms. If in pots or rings, a soilless compost is preferred.

During growth it is essential to control the watering, which is more easily achieved by the use of the ring culture technique. Do not overwater or feed, rather starve the plants. The flowers will be very poor, but it is not good flower spikes that we are after. The best, well-ripened seed is produced under these conditions. A starved plant gives high priority to its seed production and if the going is hard it will concentrate its entire effort on producing viable pollen, a sticky and receptive stigma and good, well-ripened seed. This is nature's way of ensuring survival of the species! As further example, the starvation process is extensively used by chrysanthemum breeders of exhibition cultivars, being the only way to obtain seed from a huge, many petalled bloom. It is also used by the Japanese in breeding their giant chrysanthemums, lilies, and double petunias. The most expensive $F_1$ hybrid seed (which ultimately produces a 5 ft (1.6 m) chrysanthemum plant) comes from a single, starved daisy-like bloom not more than 1 in (2.5 cm) across on a 7 in (15 cm) high plant. Minimum irrigation with no feeding other than correction of trace elements or magnesium deficiency is the technique used.

When the flowering spike appears from the sheath of leaves, break off all the buds except the bottom four, which will be retained to produce those valuable seed pods. It is not necessary nor desirable to use the whole spike; concentrate on just those four pods. In fact four good seed pods should give a fair quantity of seeds, and usually all that is required of that particular cross. Tie the spike to a cane, the top of which should be just about level with the uppermost retained flower bud. Just before the bottom flower opens place a paper bag over all of the buds as a precaution against bees or other insects spreading stray pollen. In a well-controlled breeding programme it is necessary to make sure of the true parentage of the seeds produced. Polythene or greaseproof bags are not good as they induce high humidity inside, making a marvellous habitat for moulds, mildew and rot fungi of various types. I make mine to the appropriate size by folding newspaper and sticking the edges with adhesive tape. At the same time, and certainly before the pollen is shed, each floret must be emasculated by removing the three anthers. This will prevent self-fertilisation. A pair of fine tweezers is very useful in ensuring that there is no damage beyond the removal of the anthers. As each floret opens, still under the protective paper bag of course, it can now be crossed (fertilised) by bringing pollen from the other selected 'father' parent and carefully depositing it on the stigma of the 'mother' parent. Some breeders use a camel-hair brush for this purpose, but it does need cleaning in a sterilising spirit between each cross and carefully drying. It is far simpler to use a finger and just wipe between different crosses to guarantee the parentage.

It is said that the best time of day to carry out this operation is midday when the pollen is ripe and the stigma more receptive, but I have had few problems in the greenhouse with the setting of seed and do the work as convenient. Incidentally, I make the same cross on all four florets to produce four pods of seeds of the same parentage. They are easier to handle and label that way. How fiddly to label four pods all different crosses — usually they set seeds and very rarely have any misses if the work is repeated over a number of days as the florets open and fade. When all four florets on the spike have faded, the paper bags can be removed to allow better air circulation and the sun to do its work of ripening.

There could be a problem if the selected parents do not flower at the same time and it is not possible to pollinate direct due to the non-availability of pollen at the particular time. Fortunately there are a number of techniques available to overcome this situation. The easiest way would be to advance or delay the flowering of the pollen or 'father' parent by cutting the flower-spike, standing it in a vase or container of water, and opening the flowers either in a sunny window of a warm room to advance it, or in a cool darkened room or even a cold store to retard it. Your friendly butcher or shopkeeper who has these facilities, or ideally a florist with the specialised equipment and correct humidifier, could help in this respect. It is surprising what can be achieved with a friendly word, and a few spare spikes as a gift can work wonders!

The technique that has recently claimed much attention is in the use of stored pollen. It has been found that pollen will store dry and remain viable for some time. This is of great importance whether you are a lone breeder or a material-swopper. Use of stored pollen enables a hybridist to combine late and early cultivars in crosses without having to resort to forcing plants to flower prematurely or even out of season. It also opens the door to 'swops of pollen' between co-operative breeders and members of syndicates. Pollen is frequently even sent through the post in suitable phials or containers. I have even 'stored' pollen for some time on flower-spikes kept indoors on which the flowers have withered and died. Kept dry, the pollen can be shaken out when required. There is also the method, frequently employed in the USA, of taking the 'father' parent indoors and collecting the pollen as it falls onto the lip petals and using this on cottonwool 'baby bud stick' swabs to pollinate or to store. Pollen collected in this way is no less viable than pollen direct from anthers. Experiments are also taking place in the cool refrigerated storage of pollen; it may be possible to establish a kind of pollen bank!

The activities of 'pollen stealers' have frequently been observed at flower shows, their equipment consisting of prepared matchboxes containing cotton wool and suitable labelling. Certain individuals are notorious for this practice and much pollen has even been transferred internationally. It is a common practice to acquire spikes of new and special cultivars at the termination of the show for the same purposes. As to whether this is ethical or not is a matter of opinion, but if spikes are taken it is only courteous to ask the exhibitor first; after all he may want to take his own spike home for the same purpose. The authors would never resort to such devious practices!

Having made the crosses — once again this is emphasised — you must record carefully the parentages. A small plant label fixed with a twist-it tie is ideal and should from now onwards accompany the seeds and plants at all times until the seedlings flower and selections are made. It is universally accepted by plant breeders that the mother parent is named first, × father parent last. All that is necessary now is to spray against pests, especially greenfly, red spider and thrips and to allow the pods to swell, turn brown and ripen, then to clean and store in preparation for planting in the spring.

Care must be taken to harvest the pods just before they split to reveal the valuable contents. Gladiolus seeds look like cornflakes with hard centres! The spikes can be cut at this stage and placed each in a paper bag, with its label of course, to catch the seeds as they are released. Place in a warm room indoors and soon they will be ready for cleaning and storing. Cleaning gladiolus seeds is easy, and a sharp razor blade will assist in detaching the seeds from the pods. Store the seeds in envelopes or small boxes in a cool dry place. The seed does not need to be sown immediately in the following season and it is always a good 'fall-back' to have saved some of your valuable and expensive material in case of accident with your initial sowing. Provided storage of the seed has been in paper, foil or tissue bags and not polythene, in cool dry conditions, it has been known to store successfully with little diminution of viability for up to five years.

## Seed Sowing and Growing

There are many ways to sow and grow on gladiolus seeds, and we shall describe ours. A start can be made in the greenhouse when danger of frost has passed, the daylight hours increase and there is the first indication that spring is on its way. There is no point in starting too soon, however. Although not necessary, a little heat just to avoid frost is an advantage. High temperatures are not desirable and can indeed be harmful as we have learnt to our cost, so shade from hot sun. There are occasional clear, sunny days in early spring that can make temperatures in the greenhouse soar to extreme conditions indeed, which could be fatal to those germinating seeds.

Prepare large pots, preferably clay 8in (20cm) or upwards in diameter, with a soilless compost. An alternative can be ring culture using anything easily obtainable as containers, old buckets with the bottoms removed, paint cans, plastic containers — the range is endless and they should cost virtually nothing. Fill the pots or rings to within 3in (8cm) of the top. Seeds can then be placed at that level on the compost. You may be surprised by that depth of sowing but we can assure you it works, and well. Actually it is not as strange as it may seem. The seeds of many of the South African and Asian species whose normal habitat is on river banks are frequently covered whilst dormant by flood waters and silt. As the floods subside the silt and grit is deposited on the seeds often to a depth of 2–3in (5–8cm). The fact that these species still exist after thousands of years is proof enough that the seed depth of 2in (5cm) is not so bizarre as may appear.

Do not allow the compost to dry out at any time, even for a very short period, and it is essential to shade from the sun with newspapers or similar Gladiolus seeds will not grow if subjected to dryness at the time of germination. You can now see one of the

81

*Figure 11 A parentage chart of gladiolus cultivars*

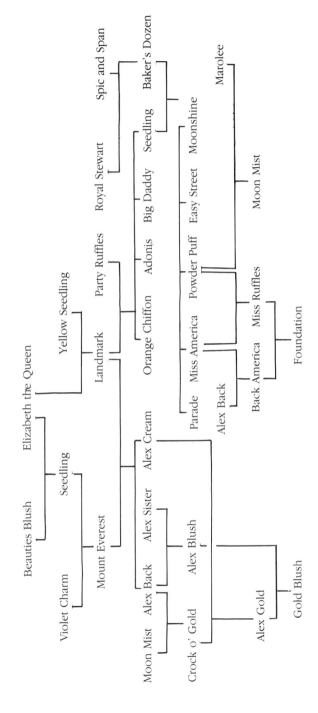

*Note:* some cultivars in the later generations are not yet released, the names are still "stud" identification names.

reasons for the deep planting: moisture is more even at that depth. Germination should be excellent by this method. In the wild most gladiolus seeds germinate in shade in grass or other plants and therefore are protected from strong sunlight in the first season. It is best to remember that to grow any cultivated plant really well it pays to study the environment and habitat of the species in the wild.

Some growers de-wing the seeds before sowing. It is claimed that the germination rate is increased and that by keeping the kernels only, they are easier to handle. We never de-wing ours. The wings are provided by nature to aid transportation by the wind and we cannot see that removing them can have any advantage whatsoever. If the seed is to be kept and not planted immediately we consider it to be a disadvantage, as de-winged seed does not store nearly so well as natural seed.

# *Chapter 8* **Gladioli of the Future**

## Into the 1990s

If the gladiolus enthusiast of today were asked to study objectively the 'state of the art' on the gladiolus scene globally or internationally, he would find it difficult to see any really major unfulfilled requirement in the gladiolus specification. The past 15 to 20 years have given us small but regular improvements without any spectacular break-through since the inter-species fragrant hybrid of new Zealand, 'Lucky Star'. Step by step the range of colour, the variation of the floret size and type, and the health of the garden gladiolus has been improved.

We now have the 'all year round gladioli' in the sense of being able to produce florists' gladioli at any season or month of the year. Nothing has been done to the cultivars, but scientists have come up with programmes for long-term storage or the early dormancy break of mature corms so that corms can be prepared to accommodate any planting date. Soil conditions, daylight lengths, temperature and humidity levels can now be controlled by computer so as to produce flowers from corms 75, 90 or 105 days after planting. The flowers so produced are uniform in size and flowering date. In the future what is already a *fait accompli* for chrysanthemums and freesia will be common practice for commercial all-year-round gladioli.

Cultivars that have proved to be highly successful in the market place will be tissue cultured from selected stock, emphasis being placed on virus-free clonal propagation giving high multiplication rates. Grown hydroponically in sterile media under controlled temperature, humidity and light conditions, the mass production of flowers and corms of good gladiolus will be, in the near future, an economic proposition.

The advances in science and technology over the past decade in such things as the electron microscope and microphotography have given to the hybridist and botanist the tools to obtain undreamed of information in important fields like chromosome numbers and pollen compatibility. In other fields of technology we have the means to identify readily, and quickly, the pigments responsible for the colour of flowers, and in the nutritional aspects of growing it is now possible to follow the flow of nutrients and measure the needs of plants element by element. By the same methods the efficiency of systemic pesticides or growth-regulating hormones can be monitored. All these new advances point the way to better versions of our current favourites in the near future.

In the 1970–80 period some growers were critical of hybridists in their failure to produce a good, healthy, true-blue cultivar, whilst other critics pointed to the lack of fragrance in the summer-flowering gladioli despite having fragrant ancestors. The lack of true-blue cultivars in the gladiolus is due to the absence of the pigment (plastid) delphinidine in the genes. Nevertheless, by measuring the blue/red ratio that the colour agent malvidin has imparted to a hybrid, the raiser can now be guided as to the 'bluest' of his seedlings. Line breeding from high blue-ratio purple and violet gladioli gives almost true blue flowers. The year 1988 saw a new strain of very good, strong, tall and healthy 'blue' gladioli of exhibition standard. Only a few years more and the true-blue gladiolus will arrive.

Basically the same approach is being applied in the production of black-red gladioli, though here the problem is not in the intrinsic chemistry of the flower. Black-red gladiolus flowers have very little reflective power, and it does not need an advanced diploma in physics to know that black or black-red surfaces are therefore much more absorbent of heat than shiny surfaces of lighter colours. Many of the black-reds and deep black-violet gladioli of the future will be magnificent. Growers will be advised to pick the flowers as the first floret bud opens, then using a nutrient solution in the water allow his blooms to expand in 10 hours per day of subdued light, with some ultra-violet content, of course. Soon we shall see these new creations big, black and beautiful!

Not far ahead now is the fragrant summer gladiolus derived from *Gladiolus callianthus* (formerly *Acidanthera bicolor*). A stable tetraploid strain of *G. callianthus* has now been created. This species variant will give fertile crosses with any of the numerous tetraploid garden hybrids. At least 50 per cent of the seedlings have some fragrance and about 5 per cent are as fragrant as *G. callianthus* itself. Work on these fragrant hybrids is being done in Britain, California, Czechoslovàkia, New Zealand and the Soviet Union. Pink, red and rose-coloured fragrant hybrids already exist, as well as the white/maroon blotched version from work in the 1970s. The future prospects for a coloured, fragrant gladiolus cultivar of commercial value are very positive indeed.

Sadly, one cannot rate very highly the chances of producing winter-hardy gladioli of acceptable quality. Little progress has been made and there are very few encouraging signs. The species hybrid *nanus* group has been used extensively in the search for a garden cultivar to grow in winter-frost areas. Some of the earlier *nanus* cultivars are semi-hardy in that they will safely over-winter in sheltered positions where the ground temperature does not fall below −5 or −6° (24°F). This constraint limits the areas where they can be grown.

It was firmly believed that if *nanus* hybrids could be crossed with some of the Eurasian species, the hardiness of the resultant hybrids would be sufficient for survival in European winters. There are however two problems. Firstly, the readily available species from Europe, Turkey and Iran do not readily pollinate with the *nanus* hybrids, and those pods which do survive seldom produce viable seed. The second problem is the very slow propagation rates of *nanus* × Eurasian, either by seed or offsets. Not so much a problem as a disappointment is that seedlings finally obtained are usually in shades of dull purple and certainly no improvement in size or form to either parent. There could be a future for this Eurasian × *nanus* hybridisation if it becomes possible to get pollen from hybrids of matched or compatible chromosome number, with the resultant seedlings meristem-tissue cultured. This idea is still an idea, but perhaps the future will enhance it to a practicality.

In general terms the future of the gladiolus is good, the flower itself is becoming known worldwide, (300 million people must have seen the gladioli bouquets awarded to medallists at the Seoul Olympics 1988), and is being grown in many countries. The gardener is having his gladiolus requirements met in quality, health and cost terms, and even the strongest requirements of the exhibitor are fully met, though at some extra cost. As cultivars deteriorate they will be replaced by better and healthier versions; in that respect the future is very good as we have more good hybridists than ever before. The range of types, sizes and colours of gladioli are being expanded year by year.

There will always be hybridists striving for something new, different and better. So long as they remain active and their numbers can be matched by sufficient gladiolus lovers willing to compensate the hybridist by buying and growing the new introductions (albeit at a somewhat higher than usual price), the gladiolus will remain a favourite summer flower. The future of the gladiolus is largely in the hands of those who grow them, sell them, buy them and breed them.

The plants are there; the future now depends on publicity, information and marketing techniques. The seventeenth century flower must be presented to the world of the twentieth and twenty-first centuries reminding us all that 'a thing of beauty is a joy forever'.

## Breeding for Fragrance

In pursuing the search for fragrance in summer-flowering gladioli perhaps the most serious and progressive hybridist in the world today is Igor Adamovic of Bratislava in Czechoslovakia. Adamovic had the foresight to tackle the problem of producing fragrance from two distinct angles and this has proved very successful. Despite many setbacks over the past few years he has now the most progressive assortment of fragrant cultivars of his own hybridising.

The mainstream parents of his current fragrant cultivars are:

1. A chance fragrant seedling produced by the late George Webster of USA and recorded by George as **S-4 (scented no.4)**

2. **Fragrant Beauty**, a cultivar from Spencer (USA) 1957

3. **Yellow Rose**, a cultivar also from Spencer (USA) 1952

4. **Acacia** a famous fragrant from Rev C. Buell (USA) 1956

5. **Bouquet**, from Rev C. Buell (USA) 1963

6. **Angel's Breath**, from Rev C. Buell (USA) 1970

7. **Lucky Star**, the *Acidanthera* × 'Filigree', the first ever gladanthera, by Mrs Joan Wright of New Zealand

8. *G. callianthus* species, *var, 'bicolor, Murialae'*

9. The tetraploid version of No. 8 above

Whilst using the above material to make crosses with lightly fragrant or non-fragrant cultivars, Adamovic early on established that the genes for intensity of fragrance were mainly recessive and that heavy fragrance did not always come out in $F_1$ generation but required inter-crossing and selection for fragrance.

Adamovic has given me the following parentage information from his extensive and meticulously compiled records. These records are kept on a Sinclair 48K computer which he has maintained himself despite a multitude of problems:

'Honey' (in Slovak language 'Med') — 'Acacia' (Buell '56) × 'Chiquita' (Pierce '59)
'Alpha' — 'Honey' × 'Inca Chief' (Fischer USA)
'Albeta' — 'Alpha' × 'Beta'
'Broskynovy Kvet' (Kvet=flower) — 'Heritage' (Greisbach) × 'Fragrant Beauty' (Spencer '57)
'Callianthea' — 'Honey' × *G. callianthus (A. bicolor)*
'Citronovy Med' (Lemon Honey) — 'Yellow Rose' (Spencer '52) × Med
'Eau Sauvage' — 'Dream Street' (Greisbach '64) × 'Lucky Star' (Wright NZ)
'Malinovka' — ('Acacia' × 'Aaron's Exotic Orchid') × (S-4 Webster × 'Med')
'Medovina' (Honeyed Wine) — 'Honey' × ('Lucky Star' × 'Bouquet')
'Jaraslava' (79) —'Lucky Star' × ('Burma' × 'Chantilly Lace')
'Pokrok' (Progress) — 'Heritage' × 'Angel's Breath' (Buell '70)
'Felicia' — 'Acacia' × 'Angel's Breath'

In all the Adamovic cultivars there is a persistent daytime fragrance for the first 48 hours of each floret, and though fragrance varies from hybrid to hybrid the main characteristic of the fragrance is 'floral', that is the observer invariably equates the fragrance with a similarity to other notable fragrant flowers such as rose, jasmine, hyacinth, honeysuckle etc.

The future for this now established strain of summer-flowering gladioli is to select for fragrance and attempt to link to the fragrance longer flower-heads, higher bud counts and, in the case of the seedlings with *G. callianthus* parentage, an improvement in the floret shape to take it away from *G. callianthus* towards the more formal floret shape of the normal summer-flowering gladiolus. A number of the fragrant seedlings are frilled and ruffled, with an attractive waxy finish to the petals, and would be acceptable gladioli even if they had no fragrance. Propagation is not good as they produce few cormlets, a problem frequently encountered with hybrids from such multi-source parentages.

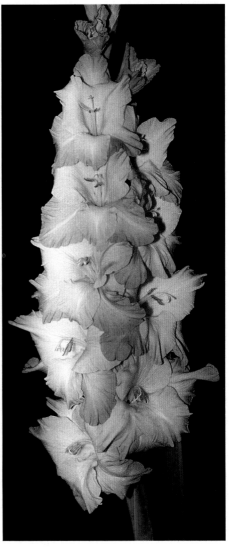

1. 'Vicki Lin'. A large-flowered exhibition cultivar.

2. 'Parade'. The number one large exhibition gladiolus of all time.

3. 'Moon Mirage'. A frequent Grand Champion from the Giant size exhibition classes.

4. (Left) 'Moon Mist'. A fine example of the ruffled floret in exhibition gladioli.

5. (Below right) 'Jester'. An exotic sport with laciniated petals.

6. (Bottom) *Gladiolus nanus*: a spring-flowering species hybrid, 'Amanda Mahy'.

7. (Top) a prize-winning selection of florets of primulinus hybrids

8. (Bottom) a champion primulinus hybrid seedling, 'Rutherford'. (A.W. Brown, Lancs.)

9. 'Amy Beth', an outstanding show miniature in a lovely colour combination.

10. 'Lynn', one of the best red show miniatures available. (Raiser; Walker, USA)

11. 'Winsome', a reliable and formal show miniature. (Raiser: Roberts, USA)

12. 'Pegasus', an easily grown small-flowered (100 class) show gladiolus. (Raiser: Visser, Netherlands)

13. (Below left) 'Tendresse', an outstandingly successful show gladiolus in the medium (300 class) size. (Raised in the Netherlands)

14. (Top right) a prize-winning trio. Left to right: 'Carrara', 'Moon Mirage' and 'Grand Finale'.

15. (Bottom right) a championship basket display.

16. A top-rated Canadian exhibition cultivar, 'Drama'. (Raiser: McKenzie, Ontario)

*Gladioli of the Future*

By inventing new and modifying established propagation techniques, Adamovic has made spectacular progress with the fragrant gladioli. In the not too distant future when a wider range of colour and form of these exciting cultivars is available, it is envisaged that these creations will be developed by meristem-tissue culture, hopefully making them internationally available and compensating the originator for his many hours of dedicated pursuit of fragrance and his innovative ideas.

## *Chapter 9*   **Recommended Cultivars**

The vast range of gladiolus cultivars now available makes the task of recommending 'the best' almost impossible. It depends on the personal taste of the grower as regards size of flower required, the purpose for which the gladioli are to be grown, and above all the location and nature of the garden in which the gladioli will be grown.

Some growers may wish to purchase and grow gladioli with a specific objective in view; others may wish to embrace all three of the main uses of the gladioli, which are — not necessarily in order of quantity — so used:

1. As cut flowers for the home, in arrangements in association with other flowers, or for competitive floral-art exhibits in local or national shows;
2. for garden decoration or landscaping;
3. for exhibition-growing for competitions in local, regional or national gladiolus shows.

In making recommendations the cultivars will be listed in such a way as to indicate whether the cultivar is better suited to one or other of the three groups indicated above.

A useful point at which to start is with the section known as *nanus* or species hybrids. These are derived from inter-species hybrids of mainly South African species, and retain the grace and charm of their progenitors. The size of the plant is about 2-2½ft (60-78 cm) overall, and on the shorter but very pliable stems the *nanus* carry up to seven florets, the individual florets being some 1¾in=2in (4-5 cm) across. This delicacy of growth and pliability of stem makes them ideal for arrangements in floral art, or indeed for home decoration in modern dwellings. The *nanus* group are also excellent garden subjects for a shaded or partially shaded perennial border. Since they are not too tall and require no unsightly supports, they fit very appropriately into the perennial border.

### *Nanus, Colvilleii*, Charm, and Summer Cultivars

In milder areas where ground temperatures do not fall below −6°C (22°F) at corm depth, i.e. 3in (8cm) deep, the plants can be left in place over winter for 3 or 4 years before there is need to lift and divide the clumps. Most *nanus* cultivars will survive normal or average British winters in most areas except very exposed regions of the United Kingdom. As a precaution the location of the clumps of *nanus* can be covered with bracken, peat or compost for the hardest winters, removing it in early March and sprinkling slug pellets around to protect the emerging shoots.

Modern storage techniques and cultural adjustments have now enabled growers of *nanus* on a commercial scale to have flowers of these delicate beauties from May to September. Corms planted in mild areas in October or early November will break through the soil about early March and come to bloom about mid-May to early June. If a range of cultivars is chosen the spring blooming period can be extended to late June.

## Retarded *nanus* for Summer Flowering

Corms of *nanus*-type cultivars lifted whilst dormant can be stored in controlled conditions of temperature, humidity and light so that the break of dormancy is retarded for up to 4 or even 6 months. If taken out of storage in March (the normal planting time for summer-flowering gladioli) and given an appropriate change in environment, the corms can be induced to break dormancy and produce the shoots and root nodules quite quickly in time for a safe planting in April to early May. Some *nanus*-type cultivars are treated as described above and can be purchased ready to plant, complete with shoots and root nodules, in late February, or even with the summer-flowering gladioli as late as April.

Corms of *nanus* arriving in February need not be planted immediately if the flowers are required for August show purposes. Corms should be stored in dry peat facing upwards in a growing position in trays in full light. Corms so treated will be in perfect condition up to early May, though by then the green shoots may be 1–1½ in (2.5–4 cm) tall. These corms may be planted at the same time as the summer-flowering gladioli and will give flowers as good as if not better than those planted in late autumn. In a normal season the March-April planted *nanus* will bloom about mid-July, and in the (hopefully) better weather of the summer months will produce larger, brighter-coloured and longer-lasting flowers.

The botanical origins of the *nanus* group of gladioli are given in Chapter 2. The following varieties are normally available, through suppliers, from Holland or the Channel Islands. Reference to Appendix III will give the names of specialists who can supply retarded corms for summer flowering.

## List of Cultivars

**Charm hybrids (Ramosus):** small-flowered type, florets about 1½in (3.5 cm) across without darts or blotches of other small *nanus* types:

'Comet' — self cherry red
'Charm' — purple-red with creamy-white throat
'Rose Charm' — light-pink with large cream throat
'Robinetta' — deep cerise-pink, lighter in throat

*Colvilleii* **hybrids:**

'The Bride' — pure white, tubular florets somewhat upwards facing
'Blushing Bride' — similar but light-pink in flush

*Figure 12  The Charm hybrid* 'Ramosus'

**Nanus hybrids:**

'Amanda Mahy' — salmon-pink with 'darts' on the three lower petals. 'Darts' are cream and carmine
'Impressive' — salmon-pink with bright-red darts
'Nymph' — pure white with crimson darts
'Elvira' — a larger version (2in/5cm) floret of 'Impressive' with darts of bright orange-red
'Guernsey Glory' — the best for summer flowering. Florets 2½in (4cm) in bright, glowing orange-red with small white mid-rib flecks
'Spitfire' — blood-red with white darts

There are no classes for *nanus*-type gladioli in the regional and national shows of the British Gladiolus Society. Some local shows will accept them in 'small flowered' classes. *Nanus*-type cultivars feature quite frequently as prize winners in the floral-art classes at regional and national shows.

The *nanus* and *colvilleii* hybrids are perhaps better known as components in bridal bouquets, corsages, and floral tributes of many kinds. In conjunction with freesias of complimentary colour these small gladioli create a very appropriate and exotic-looking accompaniment at any special occasion. I have been pleased to advise the use of *nanus* gladioli (and even supply them) 'Amanda Mahy' to functions as corsages for the ladies in place of expensive orchids. Recipients are invariably surprised to learn exactly what flower it is they are admiring.

## *Primulinus* Hybrids

The *primulinus* hybrid gladioli, known affectionately as 'prims', are now an almost exclusively British cultivated varietal group, and can be difficult to obtain even in Britain. Details of the genetic make-up and evolution of the *primulinus* hybrids are dealt with elsewhere in the book. Suffice it to say that those available cultivars finally recommended can be used as garden decoration in borders and make graceful and delicate cut flowers in arrangements. However, the cultivars suggested are also suitable for showing in gladiolus exhibition classes specifically for primulinus hybrids. Most local shows still retain a class or two in the schedule for exclusively *primulinus* hybrids (as opposed to miniature-flowered, *primulinus* or non-*primulinus*) conforming to the British Gladiolus Society requirements for *primulinus* hybrids. The BGS National Show has several classes for *primulinus* including 6-spike, 3-spike and single specimen blooms. This or a similar list of classes also feature in the major regional shows in Britain.

The requirements for a good *primulinus* are that the bloom must be on a strong but pliable thin stem, carry at least 17 buds of which preferably 5 should be fully open, a further 3 should be showing colour, with the remainder forming a straight well-spaced tip of adequate length to give balance overall to the stem. The florets must be well attached, facing forward and in step-ladder formation alternately on the stem, each floret just clear of the floret above or below it. The floret form must conform strictly to the requirements, the petals must be evenly matched for size, the wings,

*Figure 13 A perfectly balanced and presented exhibition primulinus spike*

the lips, the hood and the 'falls' being exactly as specified and consistent, floret by floret. Any throat markings (blotches, darts, flecks) must be consistent floret by floret. It is not a defect to have only one or only two petals blotched or marked in each floret, provided the feature is wholly consistent. Hoods must always be consistent, drooping to half the depth of the floret and just covering the stigma and anthers.

There are very few cultivars which regularly produce exhibition-quality spikes, but if the flowers are not required for show purposes then 'mixtures' of *primulinus*-type cultivars will give a very satisfactory range of colours suitable for floral art or home decoration. The following list is compiled on the basis of winning exhibits at major British shows over the past five years:

**'White City' (Holland)** — a good, strong-growing white with adequate bud count; may grow too large to be 'graceful' if over-fed
**'Leonora' (Holland)** — mid-yellow of good floret form. A rather heavy stem that needs staking early to keep straight; don't overfeed
**'Essex' (Holland)** — a consistent winner in bright-red. Use medium-sized corms (1 in, 2.5 cm dia) or if larger sizes do not 'de-eye'
**'Anitra' (Holland)** — blood red, good placement and bud count but occasional displacement of hood
**'Atom' (USA)** — light-scarlet edged silvery yellow; very graceful and delicate
**'Hastings' (UK)** — light beige-brown, good formation and floret form
**'Cafe au Lait' (UK)** — slightly darker than 'Hastings' with small chocolate marks on lip petals; small flowered
**'Rutherford' (UK)** — a new cultivar. Brick-red, heavily veined and laced silver; a classic show prim

## Australian Cultivars

In the past two or three years it has been possible to obtain either directly or via Leonard Butt Gladioli in Ontario, Canada, a few of the exhibition cultivars of Australian origin. The limited distribution in the northern hemisphere of these gladiolus cultivars, coupled with the infrequency with which the Australian cultivars appear on the show bench, has meant that they still remain to most growers something of an unknown quantity for exhibition in Europe and North America. I have seen and grown quite a number of Australian gladioli over the years, yet there are very few indeed that could be recommended with confidence as would-be Grand Champions in North America or Britain — as the judging rules stand at the moment in these countries.

The positive features that Australian gladioli have as regards their show-winning potential is the high bud count, long flower-head and robust growth. Mechanically the 'big Aussies' are very good in that their attachments are usually good, they travel well, manipulate easily and withstand high growing temperatures without the 'crooking' of the stem. In fact Australian gladioli win most of the Grand Championships in Australia despite very intense competition from

USA and Canadian originations. The gladiolus societies in the various states of Australia use the North American Gladiolus Council's classification for size and colour in exactly the same way as do the societies in the northern hemisphere. The Australians, however, organise their show schedules in a quite different way, and the judging rules and pointing system at shows are specifically designed to cover the range of cultivars bred and raised under Australian conditions. The major differences are in the points awarded for 'placement' and 'floret form', and the application of three separate and distinct types of flower definitions within the grandiflora section of the schedule. The spikes are designated as 'formal,' 'informal' or 'decorative', and there are cultivars that are classic examples of each. Only the formal types are comparable to the northern hemisphere show gladioli, and the requirements for a formal spike are identical, i.e. double-row placement, rectangular-flowering floret profile, slight overlap and slight taper to flower-head. Florets with consistent double lips are an important requirement.

The informal type is not limited to double-row placement or double-lip floret but may be either 'step ladder' or 'one above the other' – ribbon placement. A number of informal cultivars are grown and exhibited, these by northern hemisphere judging and pointing rules would be considered to have defective placement and balance in that they are tall, and have large tips and numerous green buds. These informal cultivars tend also to be plain petalled rather than ruffled, and invariably have single-lipped florets.

The decorative, or intermediate-type, cultivars are judged against a less stringent pointing system in which the overall beauty, floret form and colour appeal have the emphasis. Provided the flower is not grotesquely out of balance and is well grown, the placement can be either double row or step ladder and the florets plain, blotched, ruffled, fluted etc. There is no stipulation about single or double

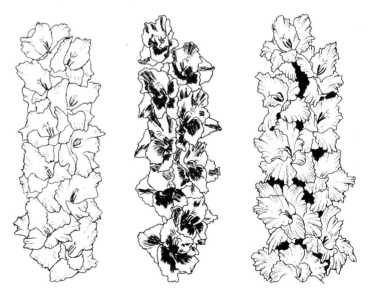

*Figure 14 Australian exhibition cultivars. Left to right: formal, informal and decorative spikes*

lips. The exhibitor perfectionist may think all this is a bit heretical, but it leads to very well-supported shows and a really wonderful range of cultivars at every major show in Australia. I do not take sides on this — it is a matter exclusively in the domain of the Australian gladiolus societies in the various states. My reason for elaborating on this interesting variation is to help you decide, should you be so thinking, which Australian cultivars would possibly be worth exhibiting in the USA, Canada or Britain.

For those who wish to dabble and 'test the water' I have listed a few names of cultivars that are well worth growing, and some of which have, to my personal knowledge, won prize cards at regional and national shows:

'Ethereal Glow' (545) Ritchie 87 — formal show-spikes 23 buds, 9 open. Light rose-pink with white throat; lightly waved petals, consistently double lipped

'Cleveland' (530) Blanden 81 — formal placement 25–30 buds on a very tall stem, making it possible to open 9 and still stay in balance. Heavy substance, nicely ruffled florets; colour apricot-pink with white throat.

'Donna Maria' (420) Podger 79 — formal placement, 25 buds, well-balanced stem. Pick the spike early and keep it cool otherwise it will open too much (10–12 florets). Faultless mechanics and an easy 'dresser'; deep gold with a pink flush.

'Extremity' (435) Podger 79 — this I believe to be the ultimate in genetic engineering. The specification was done by computer to achieve maximum show points and get into the *Guiness Book of Records*. Huge plants like tree trunks over 80 in (2m) tall; 35 buds, and it will open 14 before the earliest bud fades. Even so, it stays in balance and the placement is good. Colour is salmon-pink with a white throat. Use a 1 in × 1 in (2.5 — 2.5cm) stake and lash it straight using raffia; hire a van to transport it, then photograph it with a wide-angle lens on your camera. Not just extreme but also expensive.

'Waveney' (410) Podger 80 — formal placement, 29–31 buds, long, tall stems. Rich creamy-white with pale-yellow throat. Florets well attached and of heavy substance, lightly ruffled. Pick early, store cool and do the dressing early. Needs a bit of manipulation to achieve perfect floret positions.

If you wish to try the informal or decoratives (which incidentally are excellent for the floret-box classes at the big shows in Britain) I can recommend the following as different and beautiful:

'Dark Moon' (478) Podger 87 — velvety deep purple with a silver edge to each petal. Intensely ruffled, medium-sized spike.

'Perfect Match' (444) Ritchies 87 — rose-pink shading to blush-pink and creamy-white in the throat. Strong texture, superb form and deep, consistent ruffling; 15–17 buds but the florets will open up after cutting, right down to the last one.

'Pride of Erin' (404) Podger 87 — a deep-rich heavily ruffled green. Opens well from bud, and can hold 6 or 7 open at one time. Tall stem with well-placed florets. Will be a good arranger's flower.

'**Dark Night**' (387) **Podger 80** — a striking colour combination of deep black-violet (jet black when in bud) with a lime-green throat; florets lightly ruffled on tall, slender spikes.

'**Royal Lace**' (410) **Podger 79** — deep-cream, ruffled and heavily textured florets on a long stem of over 20 buds. Superb for corsage work because of the long-lasting florets.

These cultivars are essentially for display or floral-art use. The intense ruffling and heavy texture of the petals gives the flower an exotic, almost orchid-like appearance. Though the florets last fresh-looking rather longer than more formal gladioli, they do have a rather sad bedraggled look once they fade, so remove them at that stage. The stem will then retain its orchid-like appearance.

## Grandiflora Gladiolus Cultivars

As we now approach the vast area of choice of gladiolus cultivars may I ask the indulgence of the perfectionists and pedants if I group under the *Grandiflorus* umbrella all those cultivars that are not species, inter-species hybrids or *primulinus* hybrids. Controversy still rages amongst gladiolus fans, in particular the exhibitors at major shows, as to the correct nomenclature for the 'size regulated' classified cultivars. The key question is invariably: 'Can you reason-ably describe as *Grandiflorus* a miniature-flowered cultivar (florets under 2½in (6cm across)?' Some schedules bow to logic and describe them as 'Miniature-flowered non-*primulinus*', but this only compounds the dilemma in the view of many.

I shall take the bull by the horns (or the gladiolus by the spike) and describe the cultivars we discuss and describe as *Grandiflorus* 500 size etc., from giant to miniature. By this process and by judicious choice of illustration I hope to convey to you exactly the size form and colour of the cultivar in question.

So, let me take the plunge and begin to 'mark the card' for the giant and large-flowered section of the gladiolus cultivars (*Grandiflorus* 500 and 400 size):

Giant — florets more than 5½in (13cm) across
Large — florets more than 4½in (11cm) but less than 5½in (13cm) across.

There are many gladiolus growers and exhibitors who regard the big ones as the only 'real' gladioli and would never deign to grow anything else. To see a show bench filled with the entries in the 'Giant and Large' section at a major show inclines even the most blasé onlooker to agree 100 per cent with the 'big boy brigade'. The six-spike class at a national show is really the peak of achievement for any gladiolus grower. To select the one and only Champion Spike from these exhibits represents the most formidable task any judge will ever face. I see the Champion Spike at a national show as a triumph for three people — the person who originated the cultivar, the exhibitor who had the flower at the peak of perfection at the very hour of judging, and finally the judge who had the courage of his convictions and the experience to detect that here was the best example of the gladiolus on that great day.

The public to whom the gladiolus is just another flower can only gaze in amazement and admiration at these majestic and impressive spikes. The large and giant cultivars are essentially show gladioli. They respond to the highest calibre of care and culture, and when prepared and dressed for exhibition demonstrate the full potential of cultivars that have been bred and selected specifically for the show bench.

Bearing in mind that these majestic flowers are destined for the show bench, the exhibition potential of each cultivar is carefully assessed. A cultivar that is likely to win in one area may be totally unsuitable for another area by virtue of its lateness, its inability to withstand hot weather or rain, or perhaps because it does not travel well when not in water. Cultivars that win, say, in September in Scotland may be quite unsuitable for an early August show in London.

In Canada and the United States the large show gladioli must have between 18 and 20 ins (48 cm) of stem below the first floret. In Britain the emphasis is on balance (i.e. one-third stem, one-third florescence and one-third buds and tip) so that the stem length (or handle as it is called in Britain) is of particular importance. A variety with a short handle is less likely to be a champion wherever it is exhibited, however good the flower-head and tip might be.

Another consideration is the distance and travelling conditions that the spike has to endure to reach the exhibition venue. In North America it is quite common for exhibition blooms, strapped to square stakes, to travel up to 1,000 miles (1,609 km) and be out of water for 14–16 hours on the journey. The recovery rate and condition after being re-immersed in water is therefore very important. Like wines, some gladiolus cultivars do not travel well — it frequently depends on the substance (petal thickness) of the flower and the ability of the stem to retain water. Dutch cultivars tend to be thin petalled and have stems thinner than Canadian or American cultivars; thus they travel less well; even the best Dutch cultivars have a very poor track record at Canadian shows unless they are at local venues.

Perhaps the most important consideration in a show cultivar is the ease with which the florets can be manipulated when the bloom is being prepared (dressed) for the final staging. Florets may have to be re-positioned, opened or turned, and buds may have to be bent down or up to give perfect placement. Some cultivars have good natural placement of florets, others have not. If, however, a cultivar with poor natural placement can readily be dressed in to perfect positions it could end up as a better prospect. The degree of manipulation is often dependent upon the strength of the calyx (which holds and supports the floret) and the depth of the shoulder (attachment) from which the floret arises. The giant-flowered frequently have weak calyces and/or small shoulders so that at its largest expansion the floret becomes too heavy to stay in position.

The experienced cataloguer will know which cultivars have the various advantages, and the honest ones will comment on dressing, attachment, number of buds, length of flower-head, etc., all of which is very important information for the exhibitor. Added to this the supplier should indicate if the cultivar is early, mid-season or late so that a judgement can be made as to the suitability of the cultivar in relation to the date of the show for which entries have been made or projected.

The cultivars given here have been in the prize-winners lists in UK, Canada and USA over the past five years on a fairly consistent basis, irrespective of climatic conditions. Shows still need 'glads for cards' like 'horses for courses'. There are no sure winners, only short-odds favourites.

## Dutch Show Cultivars

In choosing to open the survey and commentary on show cultivars by turning our attention first to the cultivars of Dutch origin, we are not indicating any preference or merit rating. Historically, gladiolus shows have for many years been dominated numerically by excellent champion spikes of Dutch origin. Just as in Canada or USA, the cultivars which do best in the prevailing climatic conditions are those bred within the country concerned. Additionally, the cost of importing and acclimatising cultivars from outside Europe has made show gladioli from non-European sources an expensive item. The scene is changing somewhat now in two respects. Firstly, some good British show cultivars are arriving on the scene, and secondly, extensive growings are being made in Europe of the better non-European cultivars.

The situation was succinctly explained to me a few years ago by a very successful exhibitor from the north of England. 'Why pay dollars for a 'glad' when the trophy can be had for pence?' The value of this remark is crystallised when prices are compared: US champion: $2 for a top-size corm; British champion (Dutch cultivar): 10 top-size corms for £1.40. The gap is closing but on average the price differential remains quite significant.

*White and cream*:

'Amsterdam' 500 — huge white with massive growth and high bud count. In a dry season needs lots of water. Stake early and tie the stem to a flat surfaced support (about 1 in × 1 in (2.5 × 2.5 cm) cross section). Needs manipulation early to ensure perfect placement.

'Mont Blanc' (400) — a white sport of 'Royal Dutch' with the same mechanics: rather thin petalled, a frequent winner.

'Lowland Queen' (401) — a tall-growing white with a large cerise blotch. Good opener and natural formal placement; easy to dress and has good petal substance. If it has a fault it is the tendency to throw a single blotched petal rather than the required double lip throughout.

'Meteor' (410) — pale-cream to ivory, shading to lemon-yellow, deep in the throat. Precise formal placement with high bud count. Excellent opening qualities and keeps the florets firm longer than most and it is a fast growing early — excellent for the early shows.

All four of the above cultivars originate with J.P. Snoek of the Netherlands. Mr Snoek is a comparatively new hybridist whose cultivars are dominating the show-gladiolus market.

### Orange to salmon:

'Flos Florium' (also known as 'Praha') (432) — deep salmon to azalea-pink with a creamy-yellow throat. This cultivar has been a very consistent winner. Slightly heavier petal substance than many Dutch so dresses very easily and travels well. One of the best ever from the 'K. and M.' stable.

'Lovely Day' (536) Preyde — a mid-season giant in bright orange, almost scarlet. A reliable grower with good placement and strong attachment; tall with plenty of buds. Should there be hot weather or thundery rain about at flowering time the flowers will need protection otherwise 'burning' will occur, causing the colour to peel away leaving yellowish blemishes.

'Topaz' (425) Snoek — this new deep salmony-yellow is a seedling from the famous 'Peter Pears'. It is taller than the illustrious parent with more buds and better placement.

V for 'Vermillion' is also V for 'Veerle' (536), and 'Vaucluse' (536) — two contenders for the giant-flowered classes. The colour is distinct and eye catching. Petal substance is good, and despite their size the florets dress easily. May peel the colour in acid rain conditions or thunderstorms.

### Pink and rose:

'Maestro' (442) Preyde — this is one of the best Dutch show cultivars and amongst the first slightly ruffled ones to be introduced. It strongly resembles the American-type large-flowered with its heavy petal substance and thick stem. Mechanically it has everything: high bud count, formal placement and firm attachment. A good stem will open 9 florets without artificial aids, but a word or two of caution — stake it very early. Use a flat-surfaced stake, not a cane, and literally strap it with raffia to the cane at as many points as possible, and cover the tip with a paper bag.

'Rose Supreme' (461) K. and M. — a lovely light rose-pink with a big cream throat. Tall and formal on a whippy, long stem. A little thin in the petals and needs lots of water and foliar feed in the early stages to get the best from it. Superb for local shows. Take it to the show with the end of the stem in wet paper tissue or Oasis foam.

'Silver Jubilee' (542) Snoek — giant-flowered formal pink on a tall, strong stem. Give plenty of liquid feed once the bud is visible. The florets have good substance and dress easily — but beware the bottom floret, it is much bigger and heavier than the rest and requires extra support when travelling.

'Bit of Heaven' (460) Snoek — a mid-season rose-pink which will be Grand Champion many times once supplies increase. Good substance, formal placement and a nice, well-balanced stem with a long tip gives this cultivar all the mechanics a show gladiolus requires. Responds to tender loving care — talking to it doesn't help unless you are fluent in Dutch!

*Lilac, lavender:*

'Roncali (470) Snoek — a delicate, light-lavender sport of 'Royal Dutch.' Were it not for the tendency occasionally to throw a streak of either white or purple on the odd petal this cultivar could outdo its original. It is as good as 'Royal Dutch' in all other respects.

'Royal Dutch' (474) Snoek — a classic Dutch show cultivar in deep lavender, almost light-purple. Floret form is faultless and placement good. It has been a consistent winner since its introduction. It grows well without much feeding and if forced beyond its 400 size the petal substance deteriorates. The only fault with the 'Royal Dutch' (gladiolus) family is that once the floret wilts when out of water it takes longer than most cultivars to restore itself to full size again. The moral of this remark is: if you are showing these cultivars (sports) get to the show early and use Pepsi, Coke or 7-up in the water.

'Video (476) Snoek — a deep-lavender counterpart to 'Royal Dutch'. In cool conditions the colour is light purple. Formal placement on good long stems makes this cultivar easy to manipulate and dress. The florets are firmly attached and have more substance than 'Royal Dutch,' though perhaps the floret form is less than perfect, being very flat on the stem.

Before moving on to the more extensive and possibly much more important group of show cultivars of North American origin, it would be appropriate to extend the list of European show cultivars by including three excellent cultivars of British origin. Over the years a number of British seedlings have won top awards in regional or national shows of the British Gladiolus Society. It is regrettable that these cultivars have not become commercially available on a wider scale and consequently have not been seen often enough to establish a 'show winners' reputation.

Eric Anderton of Bury, Lancashire, hybridised and registered some beautiful show cultivars in the late 1970s and early 1980s, of which 'Alex Back', 'Alex's Sister' and 'Holcombe' were the best. They had everything in the way of mechanics that any cultivar required, and the only reason these cultivars did not achieve the fame they deserved was that there was no large-scale propagation of the stock. Ultimately, the corms of these varieties remained in the hands of a small number of exhibitors and it is only occasionally we had the pleasure of seeing them exhibited. No cataloguer has access to sufficient stock to offer them for sale.

There is, however, a much brighter story to tell about two other British cultivars that have not only show potential but solid successes in the 400-size show classes. The two cultivars have been raised by Mr John Pilbeam of Sussex, and there are others of similar high merit in the pipeline. These are being propagated on a commercial scale and should be available in quantity from 1989 onwards, with others following on each year thereafter.

In order to enable comparison to be made with Dutch and North Atlantic show cultivars the 'mechanical statistics' will be included in the description. Thereafter, for the final group of cultivars — the North Atlantic ones — no detailed mechanics figures will be given since all those mentioned will be capable of attaining in full the show-class requirements as appropriate to the floret sizes.

## British Cultivars

'Inca Queen' (435) Pilbeam — mid-season; a seedling from an American cultivar 'Inca Chief'. Perfectly proportioned and consistent florets in a glowing salmon colour with lemon-yellow throat and lip petals. Bud count is 25-27, of which 9 or 10 will be open simultaneously in perfect placement. The flower-head will reach 30 in (72 cm) and is perfectly tapered.

'Red Smoke' (454) Pilbeam — mid-season; bright mid-red overall with a silver wire-edge (picotee) on each petal. Makes a huge plant close on 6 ft (173 cm) tall. Averages 24 buds per stem and holds 9 open florets on the 34 in (82 cm) flower-head. Makes huge corms with few but large cormlets. The cormlets germinate well if soaked or peeled before planting.

## North Atlantic (USA and Canadian) Show Cultivars

Unlike the Dutch, the hybridists from North America have always raised in large numbers cultivars bred solely for their exhibition quality. Until very recently the best of the American cultivars were not exhibited frequently at British shows. The cost of corms, the lateness of arrival and thus late flowering time made the investment of up to $3 a corm somewhat of a gamble. Now many US show cultivars are grown from American stock in Holland and Britain, thus the prices are lower and the delivery dates earlier and more reliable. This will inevitably lead to more USA and Canadian cultivars being exhibited in Britain.

The number one show winner in USA and Canada has been for many years, since 1974 in fact, 'Parade' (536) (Larus 1970) late-season. This cultivar has everything a show glad needs to win and is very consistent if grown from large corms or even de-eyed jumbos, 2 in (5 cm in diameter). Despite its long list of Championship and Grand Champion wins, 'Parade' has two deficiencies. Firstly, it is late flowering, taking about 110-125 days from planting to flower production. Secondly, the handle (stem from ground level to first floret) is too short. In order to get the 20 in (48 cm) of stem as required by NAGC judging standards, or the one-third total length of stem for balance (⅓ stem, ⅓ open florets, ⅓ buds to tip) as per British Gladiolus Society judging standards, it is frequently, almost invariably, necessary to cut the stem to ground level thus destroying the corm. Progressively over the years good stock of 'Parade' has been difficult to acquire, and hybridists have been trying to combine the superb flower quality of 'Parade' with the longer handle and taller stem of other good show cultivars.

We begin our survey with 'Parade', its mutations and seedlings:

'Parade' (536) — late season, colour salmon-pink with a cream throat.

'Easter Parade' (544) Butt-Canada — a light-pink with very small white mark deep in throat. A week earlier than 'Parade' and a few inches taller.

'**Prime Time**' (500) Summerville — a large white sport with all the 'Parade' mechanics.

'**Cavalcade**' (444) Roberts-USA — a pink seedling of 'Parade' with smaller florets and a better balance. The neat ruffling and ease of manipulation makes it a superb show gladiolus for the novice.

In Canada between 1978 and 1982 a hybridiser of distinction, Mr Alex McKenzie, made several crosses using 'Parade' with 'Moon Mirage' (512) and had some marvellous seedlings which corrected the deficiencies of 'Parade' and extended the colour range:

'**Drama**' (453) McKenzie — early mid-season. Light-red blending to a bright-yellow throat.

'**Incomparable**' (545) McKenzie — mid to late season according to size of corms. Large corms must be de-eyed. Pink with a small yellow throat; very tall.

'**Pink Miracle**' (442) — mid to late. Clear light-pink blending beautifully to a small cream throat. Superb texture and ruffling. In cool conditions the florets have a delicate orange pink suffusion that catches the judge's eye.

'**Diva**' (532) McKenzie — late. Light-salmon blending to a small cream throat. The huge florets are artistically ruffled and well attached; the lower florets make superb specimens for the floret-box classes.

There are, of course, many very good show cultivars not connected with 'Parade' but having very good mechanics and show records in the 400–500-size classes.

Fortunately these cultivars cover a wide colour range which enables the exhibitor to exploit those classes calling for cultivars within certain colour ranges, a practice more common in USA and Canadian schedules than those in Great Britain.

Here are a few successful show cultivars in the colour ranges:

### *White and cream*:

'**Carrara**' (400) Squires — mid-season. A very reliable white of excellent floret quality and form. A frequent winner in America and Britain since 1986. This is a white sport of 'Ecstasy' (q.v.).

'**Ice Cap**' (400) Carnefix — mid-season. Consistently in the 'Top Ten' since 1984. Floret form and ruffling are outstanding. Attractive silvery-white with a crystalline finish. Does well from medium (no. 2)-size corms.

'**Miss Minnesota**' (400) Fischer — mid to late season. A beautiful ruffled, silver-white with impeccable placement and floret form. Consistently single lipped with precision overlapping. Use single-shoot, medium-size corms (golf ball size) and shape if available.

'**Vicki Cream**' (410) Squires — mid-season. A clean cream sport of 'Vicki Lin' with huge flower-heads. A great exhibition prospect that came in the 'Top Ten at number 7 within the first two years of introduction in 1984 and is still 'Top Tenning'.

*Yellow-orange and salmon*:

'Celebrity' (512) Roberts — mid-season. A giant light-yellow. This is grown in Canada over 6ft (173cm) tall with as many as 11 of its 27 buds open. A gladiolus of great substance with colour perfection. This cultivar has three times won Grand Champion at shows after having travelled over 2,000 miles (3,200km) by air. A Top Ten variety.

'Finesse' (435) Summerville — mid-season. Light-salmon with a bright-yellow throat, beautifully tapered spikes with stylishly ruffled wide-open florets. Introduced in 1987 it has already collected Grand Champion ribbons.

'Moon Mirage (512) Bebenroth — mid-season. Very large plain florets wide-open and flat on a long, slender stem. Not only is this cultivar the number one show yellow but it is an excellent pollen parent for hybridisers, imparting its classic show qualities to its offspring.

'Deliverance' (434) Walker — mid-season. If grown from large de-eyed corms this cultivar gives superb 500 size spikes of classic show quality. The bottom floret is frequently over 6in (144cm) across. Best described as a wide-floretted salmon-orange. Easy to grow and dress. Imposing and impressive sight if well presented. Draws forth comments of 'What a size — what a colour!' Probably needs to be re-classified as 534.

*Pink and red*:

'Attraction' (447) Coon — late. Has already achieved distinction in the 400-size class reaching the Top Ten only three years after introduction. A most beautiful shade of deep-pink blending to a clean white throat. The delicate and consistent ruffling enhances the beauty of this classic spike-maker. Easy to manipulate and a very good propagator.

'Allure' (442) Walker 87 — mid-season. Pastel-pink fading to blush-pink at the centre. Perfectly attached, precision placement and the acme of balance and taper. Won several championships as a seedling and will have immense show potential once it is available in quantity. Best recent introduction in USA in 1987 and 1988.

'Cornell' (454) Walker-Moffatt — mid-season. Velvety bright-red with excellent colour saturation. Resistant to rain or heat. Holds its florets well. Placement excellent and formal. Rather plain petalled but easy to manipulate. Travels well.

'Heavenly Pink' (445) Fischer — early. A seedling of 'The Queen' also an All-American Selection. The colour is a delicate and distinctive pastel-pink shading to blush-pink at the throat. Tall slender stems holding up to 12 open florets if allowed to. For show purposes cut early and hold cool to prevent over-flowering at judging time. Plenty of buds and a precisely fashioned profile.

'Poise' (447) — mid-season. Deep-pink with a creamy-white throat. Formal placement of well-shaped robust florets of good substance. Dresses easily and travels well, but needs irrigation to stretch the spikes. Good colour saturation.

'The Queen' (441) Fischer — mid-season. All-American Selection

1982. Pastel-pink with silvery-white throat. Very formal exhibition cultivar that needs little manipulation. Keep stems well strapped to stakes to prevent crooking. Opens very quickly, so cut early and store cool about three days prior to staging.

'Queens Blush' (441) Summerville — a blush-pink nearly white sport of 'The Queen'. All the good qualities of 'The Queen' in a distinctively different colour.

'Vicki Lin' (440) Squires — mid to late season. Exhibitors will love this cultivar; it produces a ribbon of beautiful pale-pink flowers on a sturdy, straight stem. Give plenty of liquid feed to stretch the stem and increase the 'handle'. Florets are wider than most 400 cultivars and can go to above 400-size limit.

*Rose, lavender or lilac*:

'Ecstasy' (462) Carrier — mid-season. Has been Grand Champion in Britain, Canada and USA on many occasions. A light rose-pink almost white when fully expanded. A fantastic show gladiolus that responds well to care and timely attention. May lack substance in a dry, hot season. Irrigate well to stretch the spikes.

'Award' (463) Walker — mid-season. Medium rose-pink blending to a white throat. Heavy substance and deep ruffling. Spikes stand over 60 in (144 cm) on very robust plants. Despite heavy florets, the attachment is good and placement formal. Travels and dresses well. The large florets make excellent floret-box entries.

'Picasso' (478) Skolaski — early. Formal rosy-purple. Ruffled florets in perfect placement on straight stems. The largest of the early 400-size cultivars. Strong petals that are easily and safely manipulated. A good cultivar for the novice exhibitor.

## Medium-flowered Gladiolus Cultivars for Shows

The medium-flowered cultivars are similar in most respects to the large and giant-flowered except in the size of the floret and the required minimum number of buds on the stem. Florets at the widest part must be over $3\frac{1}{2}$in (8.5 cm) and under $4\frac{1}{2}$in (11 cm). The bud-count requirements for a balanced show spike are: total 18, open 7, in colour 5, green buds and tip 6. Extra judges' points are given for numbers above the requirement minima.

The cultivars reviewed and described are all capable of achieving and frequently exceeding the required minima in terms of bud count. There are other cultivars which are also capable of producing show quality spikes.

In compiling the list of cultivars suitable for exhibition we have taken into account two factors which arise from the compilation of show schedules for national and regional shows of the Gladiolus Society and NAGC. The primary consideration is that there are fewer classes in the schedule for 300-size cultivars and fewer 'colour group' classes. The second consideration is that the 300 classes are normally 'any colour', thus the lighter and easier colours predominate. It is generally true that deep-purple, black-red, violet or brown-coloured gladioli suffer badly in heat or thundery, heavy-rain conditions. The two 'black' cultivars quoted are exceptional.

The scientific explanation we are most willing to accept for the colour damage phenomenon is, for black-red and deep-violet, that being black when in bud when the colour is just showing the floret absorbs the sun's heat entirely having no reflective properties. This heat absorption so dehydrates the protruding petals that they shrivel and harden at the tips, thus preventing the bud from expanding. For the other dark colours, as the intense colour is brought about by heavy pigmentation of the surface cells of the petals, this colour is affected by acid rain which etches, dissolves or changes the chemistry of the pigments, and the petals when expanded have white, yellow or grey flecks in them and in extreme cases small pin holes. It is an established fact that thunder rain is more acidic than normal precipitation due to dissolved oxides of nitrogen, adding to 'normal' pollutants.

Many of the really beautifully coloured cultivars in the 300-size range make excellent landscape, arrangement or cut-flower gladioli. Minor blemishes which could rule out as an exhibition spike an otherwise satisfactory gladiolus are readily hidden or become unimportant in a well-displayed arrangement.

Any of the following cultivars given good cultural conditions can be grown to exhibition standards. As the 300-size cultivars seldom exceed 4½ft (130cm), a stake of lesser size than that used for giant or large-flowered is quite adequate.

'Apollo' (333) Fischer — an All-American Selection and frequent show winner. A pleasing blend of orange, salmon and yellow. Beautifully ruffled and well-shaped floret.

'Antares' (356) Roberts — long flower-heads of dark-red ruffled florets on tall stems. The stems are very strong and flexible. Good weather resistance. Opens 10 florets.

'Atlantis' (387) Summerville — the best exhibition blue yet. Sky-blue with a clean white throat. Floret form very consistent and a good opener; holds 9 open easily.

'Blue Angel' (387) Summerville — early flowering. A clean deep-blue with a violet-blue throat mark. Rounded florets formally placed in a long strong stem; will open 9 florets on a 25-bud flower spike.

'Cher Ami' (374) Roberts — warm pinkish-lavender. Tall, straight stem with perfectly placed florets of good substance; travels well.

'Dianne' (375) Cox — a light silvery-purple with three distinct white median lines on the lower petals. Beautifully shaped flat florets with excellent attachment. Dresses easily and travels well; has won championships in Canada and Britain after 16 hours travelling.

'Fiesta' (327) Greisbach — brilliant-orange and yellow bi-colour. Heavily ruffled and formally placed florets on good stiff straight stems; will hold 7 or 8 open florets.

'Green Woodpecker' (303) K. & M. Holland — lime-green with maroon blotch. Plain but needle-pointed florets well attached and of good substance. Needs early staking and tying-in to ensure placement. This cultivar is best grown from unflowered cormlet-grown stock choosing corms about 1½in (3.5cm) diameter. To retain the full greenness pick the stem when the first floret opens and store in a cool, dark cupboard until ready to stage.

'High Seas' (382) Fischer — early. Cool lavender-blue shading in to a large white throat. Produces strong spikes with plenty of buds and a long, straight tip.

'Linda' (345) Roberts — early. Smooth medium-pink with a green throat. Waxy substance and heavy ruffling gives the florets a rather robust appearance. Not so tall as many 300-size cultivars but has enough buds to give a beautifully balanced stem.

'Mr Fox' (325) Fischer — early. All-American Selection. Reducing the superlative studded publicity description to more objective language we have a superb show or arrangement gladiolus. Newly opened, the florets are medium-green turning to golden-yellow as they mature and developing an orange edging. The florets are winged and ruffled. Excellent placement and bud count; holds 7 open florets easily.

'Pink Slipper' (345) Sprinkle — late. Medium-pink with large white throat. Long flower-heads on straight, strong stems. This cultivar makes excellent show spikes but has a tendency to over flower, opening 10 florets. To prevent this, cut when two florets are fully open and hold them (with packed cotton wool in the floret-tube and a loose rubber band) in a cool, dark cupboard until 4 or 5 more open. Try to stage with only 6 or 7 open florets to give balance.

'Rampart' (356) Butt-Canada — early. Long, straight, plain but perfect dark-red with claret-red throat. Perfect placement and balance with 7 or 8 open florets. Stands heat well and does not fade.

'Sabre' (356) Roberts — early. Heavily textured and ruffled deep-red with a velvet sheen. Formal placement on good stems. Some florets will have silver edges to the petals when young. Pick in early bud and keep stems in diluted foliar feed to preserve colour saturation. Balanced at 8 open florets.

'Tendresse' (344) Snoek, Holland — mid-season. A most consistent show winner with perfect mechanics. Deep-pink slightly lighter in the throat. Bud count is excellent carrying 23–25 on a good stem; 7 or 8 open at judging time will give a beautifully balanced spike. If not try 6 and get the 2nd Day Champion.

'Tesoro' (314) Roberts — mid-season. A fine medium-yellow that has established itself in the Top Ten show glads against all comers in all sizes. It is a pure daffodil-yellow with no other markings and has excellent mechanics. It is ruffled and frilled with wide open florets in precision placement.

'Violetta' (384) Walker-Moffatt — early. A plain-petalled medium blue-violet with each petal edged silvery-white. The flower stems reach 50 in (120 cm) and carry up to 23 buds. The florets overlap beautifully and with 7 open a well-balanced spike is easy to obtain. The colour is apt to fade if subjected to strong sun.

The 300-size cultivars, if given the same culture and feeding as the larger-flowering types, are likely to develop one or two secondary spikes especially if single-shoot corms are used. The preferred corms are large or jumbo size without removal of any shoots at planting time. It is best not to remove the secondary spikes until immediately before staging as this can cause the florets to go over-size. Oversize florets do not disqualify if registered and classified cultivars are involved. The larger florets do however affect the profile and balance of the exhibit and could lead to down-pointing of an otherwise good exhibit.

## Small-flowered Gladiolus Cultivars for Show

The small-flowered gladioli are scaled-down versions of the larger-flowered types and have basically the same requirements for exhibition purposes. The floret size must be over 2½in (6cm) but under 3½in (8.5cm). Bud-count requirements are, total buds 16, open 6, in colour 4, green buds and tip 6. Exhibition-quality stems of 200-size cultivars are best produced from corms of about 1½in (3.5cm) diameter with high crowns. Alternatively, larger corms may be used without the removal of any secondary sprouts or shoots. In general, small-flowered gladioli will produce good-quality spikes without much feeding provided they have enough water during dry spells. Canes of about 36in (87cm) height give adequate support to flowering stems.

The following cultivars, some of fairly recent introduction, have a record of success in major shows:

'Georgette' (225) Visser-Holland — early. A dazzling-orange with yellow throat. The best small-flowered show gladiolus ever to come out of Holland. Consistent and easy to grow. A good novice's flower. Tends to over-flower and may carry up to 10 open florets. A 21-bud stem is well balanced with 7 open florets.
'Holland Pearl' (290) Visser — mid-season. A formal stem with good placement and bud count. Plain petalled, deep reddish-brown (near chestnut) with white throat darts. Easy to handle, opens 6 florets.
'Amy Beth' (274) Roberts — mid-season. Medium-lavender with a creamy-white throat. Waxy texture, ruffled and frilled. A beautiful show glad with a great record.
'Winter Sport' (200) — a creamy-white sport of 'Amy Beth'.
'Astro' (287) Walker — late. A deep blue-violet with contrasting white spears on the petal ribs. Good placement and form. Opens 8 round florets. In bright sun or heat, shade the blooms to prevent fading of colour. Pick early and open up in a cool, dark cupboard.
'Bel Air' (264) McKenzie, Canada — early. Deep-rose lightly ruffled, perfect placing and form. Florets well attached and robust. Won a Championship in its first year.
'Black Lash' (258) Fredericks — mid-season. The best black-red ever produced and the perfect show type. Keep it shaded in bright sun or heat to avoid shrivelling of petals. Buds jet-black opening deep blood-red. Pick early and store and manipulate florets in cool, dark conditions.
'Blue Bird' (285) Baerman — early. Mid violet-blue with greenish-white centre. An All-American Selection and a good consistent producer of show spikes.
'Arrows' (201) Walker — mid-season. Perfect 200-size white. Ruffled florets of heavy substance, greenish ivory-white with light-yellow mark in throat.
'Manna' (200) Roberts — mid-season. A partner to 'arrows' and a very consistent grower. Ruffled florets of perfect size and place-

ment. Double lipped throughout, each floret identical.

'Darlene' (247) Summerville — mid-season. Pastel pink with cream throat. An improved 'Gigi' with long, straight stems and perfect placement.

'Dave's Memory' (258) Verinis, Latvia — an international glad raised in the Baltic Republic of Latvia, introduced in Canada and named after an American. A formal, plain-petalled black-red that can withstand sun and heat. Jet-black buds and very eye-catching.

'Candice' (235) Summerville — early. Should really be class 135. A dainty orange-red with scarlet lip petals. Dainty placement, zig-zag on the stem which is slender and pliable. Beauty and delicacy combined.

'Gigi' (247) Walker — very early. Has won more awards than any other miniature. Deep-pink with a small white throat. Perfect show spikes from well-grown no. 3 corms. Do not remove secondary spikes until the stem is cut for exhibition.

'Gold Nugget' (216) Fischer — mid-season. A deep golden-yellow with lovely frilling and shape. Florets identical all up the stem. An excellent show flower and easy to grow.

'Red Bantam' (254) Baerman — early. The best red show miniature, easy to grow and consistently of show quality. Develops quickly once cut. For show purposes leave it on the plant to the last minute.

'Rebecca' (299) Roberts — very early. Deep-brown edged with gold. Heavily ruffled and of good substance. A bit on the small size for a 200 class. The zig-zag placement is perfect, like little brown orchids on a twig.

'South Seas' (237) Greisbach — early. Bright-orange with coral-pink suffusion and a rose-pink blotch. A fascinating colour combination. Heavily ruffled florets in good placement on a strong stem.

'Winsome' (211) Roberts — mid-season. Creamy-white suffused rose-pink with a deeper rose blotch. A long stem with average 21 buds of which 8 or 9 will hold open. Great show potential.

'Arianne' (237) Walker — early. Deep-salmon with a feathered yellow throat. Formal placement on long stems, up to 23 buds. Can hold 9 open and still be balanced.

A number of other cultivars in this 200-size class in a wide range of colours and with attractively shaped florets are available and recommended for purposes other than exhibition. Descriptions, without the show classification code, are given under the section 'Gladioli for Landscaping, Cut Flowers and Floral Art Arrangements' below.

## Miniature-flowered Gladiolus Cultivars for Show

It is only in the show schedules of major shows that classes for miniature-flowered gladioli (100 size) are included, and in many of these shows the 200 and 100 sizes are frequently combined. The national and regional shows organised by the British Gladiolus Society invariably include classes for 100-size cultivars. There are classes for 100-size gladioli in Canada and in the USA where the miniature gladioli are frequently referred to as 'Pixiolas'. Pixiola was the florist trade name chosen in a widely publicised competition throughout the United States of America. The winner gained considerably for his invented name; sadly however, it made very little difference to the sales of the delicate little flowers.

In comparison with the medium, large and giant-flowered gladioli the numbers of miniature (100) gladioli grown is very small, probably the florist trade estimate of 2½ per cent is still valid. Those involved in the commercial production of gladioli for florists and gardeners give as the main reason for the lack of demand for miniature glads that they lack substance and travel badly. Miniature gladioli cannot stand being too long out of water and the buds fail to open satisfactorily when the stem is replaced in water after a journey. Exhibitors have overcome this problem for show specimens by transporting each stem (or up to 3 stems) in a well-soaked block of Oasis floral plastic foam, using Phostrogen or a similar foliar feed chemical as the soaking liquid.

In the early days of the 100-size gladiolus it was not uncommon for so-called 'miniature' gladioli for the flower market to be produced from large cormlets of 200 and 300-size cultivars. These were in fact under-developed stems with few buds and represented poor value for money. The show requirements for 100-size cultivars are very demanding and judges, whether British (BGS) or North American (NAGS), interpret the rules quite strictly. Particular emphasis is placed on floret size and bud count, with point deduction mandatory for deficiencies in buds and oversize florets. The 100 class of cultivar is the only one where there is no minimum floret size, only a maximum, of 2½in (6cm). The bud counts are different between British and North American. The British requirements are given first with the North American in brackets: total buds 14 (15), open florets 5 (5), buds in colour 3 (4). Additionally, the North Americans only require a flower-head of at least 22in (51cm).

The corms of miniature gladioli are usually much smaller than those of the larger flowered cultivars and the problem of oversize flowers does not normally arise if corms of below ¾in (2cm) diameter are used. If however older and larger corms are used do not remove any 'eyes' but let as many flower stems develop as possible. Do not feed via the roots but use foliar feed only up to the stage of coloured bud (i.e. about 6 days before first floret opens), and do not remove any secondary spikes until you are ready to stage.

The number of reliable show cultivars is limited, and despite having grown or seen about 60 cultivars in the last five years all of which have been launched with glowing catalogue descriptions we find only four American and two Dutch are regularly capable of producing spikes of exhibition quality.

Top of the list would be:

'Small Wonder' (143) Roberts — early. This is a truly precision miniaturised version of the huge Grand Champions. It has the formal double-row placement so rare in the smaller-flowered gladioli. The tiny florets about 1½in (3.5cm) across are lightly ruffled and at least 6 usually 8 are open on a 20-bud stem. The flower-head exceeds 22in (51cm) and the stem-flowers-tip balance is perfect. The colour is an even, light pastel-pink fading to creamy-

white inside. If only this cultivar was available in a full range of colours the 100-size classes would have a new lease of life.
The second choice is also American:

**'Firestorm' (154) Summerville** — early. A 'Red Bantam' seedling with even deeper colour saturation and an almost fluorescent sheen. Colour is bright medium-red and in cool growing conditions a very fine silver wire edge develops on the two wing petals. This cultivar has precision placement except for the bottom floret which tends to drop down (it can be easily and safely manipulated into place). It opens 8 neat florets in formal placement and the fully developed flower-head is about 24in (58cm) with 21–23 buds.

Rather newer but an excellent prospect is:
**'La Petite' (125) Klutey 1984** — late. Medium-yellow tipped orange-scarlet and a 'reflection' type blotch of light orange on the dorsal petal. This complicated colour pattern is however consistently displayed on each floret. It is tall and opens 8 ruffled florets on a 23-bud stem. The flower-head is long and the overall height is 40in (104cm). Can be a little 'gappy' though this placement is also consistent. A good 3-spike prospect for a mid-August show.

There is still a good supply of an older cultivar with a fine record:

**'Natty' (104) Dolittle** — early. Deep-green and slightly ruffled. Opens 7 florets on a well-balanced spike of 19–20 buds. The flower-head can reach 26 inches (62cm). This cultivar needs much water to stretch the stem; pick it early to retain the deep green colour and store it in the dark in a block of Oasis.

There are two Dutch cultivars that can make exhibition quality spikes if good clean stock is obtained. Unfortunately both these cultivars are prone to throw 'face-up' florets that cannot be manipulated. If placement defects are corrected early whilst the stem is on its stake a good show flower can be produced.

**'Charmer' (127) Visser, Holland** — mid-season. Deep scarlet-orange with red-brown lines in the throat. Opens 6 or 8 on a 19 bud stem. Needs some attention but can win.
**'Pegasus' (113) Visser, Holland** — mid-season. A tiny little daffodil-yellow with a red line to each petal mid-rib. Neat and formal with the classic miniature form.

## Gladioli for Landscaping, Cut Flowers and Floral Art Arrangements.

Not everyone who finds an attraction in growing gladioli is necessarily an exhibitor, and fortunately there is such a wide range of colours, sizes and types of gladioli that it is now possible to say with every confidence that there is a 'glad for every use and taste'. It is a very rewarding undertaking to produce, either for garden decoration or to cut and take indoors for vases or baskets, an armful of good, well-grown gladioli.

Many of the cultivars which are unsuitable for show make excellent gladioli for other purposes. Gladioli are eminently suited for growing in beds or borders in the home garden, or for cut flowers. The requirements for gladioli when used for these purposes is that they should have pleasant colours or compatible colour combinations, able to withstand wind and rain and to require no staking. Cultivars meeting these requirements are usually on shorter-than-usual stems but thicker and tougher. They also have fewer buds overall but open slowly and have long-lasting flowers with heavy-textured petals. It is best to avoid black-red, deep-purple, or mauve-grey coloured cultivars as these do not stand up to adverse weather like rain and hail, or to intense heat. To avoid the necessity for staking, plant the corms about 5 in (12 cm) deep in clusters of 5 or 7 corms rather than in lines or rows. The effect will be much enhanced as the patches of colour develop. For cut flowers lines are preferred. The cut-flower corms (or garden corms) can be the medium size (12–14 cm circumference) 1¼ in diameter planted 5 in deep and 6 in (15 cm) apart.

Cut-flower gladioli are best picked when no more than two florets are fully expanded. It is best to pick early in the morning. Garden gladioli should have spent florets removed every third day. The cultivars in the list are proven performers and should give a ribbon of colour in the garden for 14–20 days, and as cut flowers should last a week at least if spent florets are removed and stems trimmed back an inch (2.5 cm) every second day. Good and inexpensive cultivars include:

### Dutch:

'Hunting Song' — bright vermillion or deep orange-scarlet
'Nova Lux' — deep daffodil-yellow
'Applause' — early flowering cherry-red with deeper-red throat mark
'Green Woodpecker' — lime-green with maroon fleck
'Friendship' — light rose-pink with white throat
'Deciso' — bright red
'Flos Florium' — salmon to light-orange with yellow throat
'Oscar' — blood-red
'Peter Pears' — light-orange
'Blue Conqueror' — light violet-blue with white lines
'Wine and Roses' — light rose-pink with wine-red blotch
'Mary Housley' — creamy-white with red blotch edged yellow

### North American:

'Candyman' — deep rose-pink shading to wine-red
'Carioca' — multicoloured orange, yellow and red
'Fiesta' — yellow, deep-scarlet and orange blotch.
'Jester Gold' — deep-yellow, petals frilled and laced
'Priscilla' — creamy-white, edges of petals flaked pink
'Sabrina' — greenish-yellow with small dark-red blotch
'Annabel' — small flowered, deep-rose and light-purple
'Amy Beth' — small flowered, delicate lavender and cream

'**Orbiter**' — green centre edged pale-lilac
'**Rampart**' — bright, fiery-red shaded rose-pink
'**Rebecca**' — miniature flowered, light-tan, chocolate and gold
'**Barn Dance**' — imposing fluorescent pink
'**Bread and Butter**' — white with large butter-yellow throat
'**Blue Bird**' — violet-blue and white
'**Laurita**' – ruffled cream and white
'**Green Isle**' — small flowered, ruffled beach-leaf green
'**Ruby Ruffles**' — heavily ruffled mid-red with purplish overtones

Avoid buying mixtures as these will invariably contain a fairly high proportion of inferior or old varieties. It is much better, for landscaping, to buy distinct named cultivars and to plant them in groups of five or seven corms — or more if space is available. The colour effect of clumps of the same cultivar is much more imposing, and having the one cultivar in each colour block will help to ensure that the bulk of the plants will flower at roughly the same period to within about ten days.

If the gladioli are to be grown as cut flowers for the florists' market, plant the cultivars in colour groups and choose a range of cultivars that will give a spread of flowering time. The list of cultivars given includes early, mid-season: 90 days; and late: 100 or more days after planting to the cutting of marketable flowers. For the florists' market avoid deep-purple, grey or smoky varieties and limit the number of brown or tan in the bunches. Always pick deep-red or deep-violet blooms early, as soon as the lower buds show colour, and keep them cool and dark in a room or cooler until the first two florets open. If your clientele includes churches, funeral parlours, hotels etc., make sure you have plenty of large-flowered, tall white or cream flowers for their special needs. There are so many suitable white or cream cultivars that the best for the purpose is the one that is most readily available at the right price.

Finally, we now deal with a more modern and innovative range of gladiolus hybrid cultivars that have been produced for the use and benefit of people who do not necessarily go overboard for those magnificent show specimens. We refer more to the floral artists and flower-arrangement clubs, and those flower lovers who use flowers to decorate their homes. Included in this category are also the professional florists who prepare corsages, bouquets, posies or larger floral tributes for those special occasions, both happy and sad. For that section of gladiolus lovers the hybridists have created a new range of gladioli which by virtue of their graceful orchid-like appearance, their clever miniaturisation and their exotic colour combinations have found great favour. Floral artists have their own intuitive sense of what is appropriate to the particular occasion and need no advice in this direction. Our list on p. 93 is put forward in the hopes that somewhere amongst the flowers we describe — something will be 'just it' for the occasion.

The final group of new cultivars, the novelty gladiolus hybrids (p. 94), are less readily available than the normal type cultivars, partly because they propagate less quickly and are therefore more expensive, and partly because the demand for these *avant-garde* type flowers is small and scattered worldwide. Already smaller specialist firms are taking a more active interest in these types of gladioli and ultimately they will be available — albeit in small numbers — for the gladiolus connoisseur and florist.

## Modern Exotic Gladioli

The description 'exotic' is currently applied to the gladioli hybrids of recent introduction, many of which are distinct and differ dramatically from the more formal exhibition cultivars. More specifically, exotic is used to describe those cultivars which do not conform to the existing classification criteria or exhibition standards for judging. Exotics may be of any floret size or type and of any colour or combination thereof.

Floret shapes may be winged, ruffled, fluted, laciniated, frilled or needle pointed, these all being terms used to describe the configuration of the petals. There are others where the floret itself may be considered abnormal in that the anthers may have become petalloid, creating the impression of being doubles. Another variation to both floret form and petal shape is the 'dragon' gladiolus. In this form the petals are large and narrow — up to $7\frac{1}{2}$in long (18cm) — and have 'knuckles' or 'dragon lumps' on the rib of the petal. These protrusions are usually of a different colour — frequently green — from the main petal colour. The colour combinations in exotics may be striking and unusual, and the general colour may be enhanced by veinings and stripes of a different colour, or fringed and picotee edged with a contrasting colour. In some cases the centre has a distinctive contrasting blotch varying in size from a small thumbnail-sized marking to a large patching covering half the petal area.

'Cookie' — small-flowered; ruffled brownish-tan with pink overtones.

'Oh My!' — medium flowered; a mid-red but precisely and evenly veined and striped silver. Uniquely beautiful as a basket arrangement.

'Bambino' — medium flowered; round heavily ruffled blush-pink.

'Lili' — medium flowered, orchid like flowers, frilled and fluted in a lovely combination of apricot and yellow.

'Candice' — small flowered; a graceful bright orange-red with a fluorescent glow, like exotic butterflies on a twig.

'Carved Ivory' — medium flowered; heavily frilled and laced exotic-looking flower in cream and green.

'Rebecca' — small flowered; like little orchids. Ruffled light chocolate-brown, each petal edged gold, amber centre.

'Elfin Orchid' — small flowered; rich royal-purple, each petal edged in silver.

'Tribute' — medium flowered; heavily ruffled delicate light-pink.

'Frizzled Coral Lace' — heavily ruffled and frilled light-pink. The nearest to a double flower yet.

'Bridesmaid' — superb corsage gladiolus; 4in (9.5cm)-florets of orchid shape. Lilac and cream with petals edged in shining green and gold.

'Gloriosa' — corsage gladiolus; heavily frilled, light-rose and green.

'Blue Frost' — large, exotic-flowered gladiolus; individual florets are a display in themselves. Silvery-blue with a crystalline sheen.

**'Gold Lace'** — Primrose-yellow with frilled edge in deeper yellow.

**'Contessa'** — heavily frilled; light-rose and cream.

**'Lilac and Chartreuse'** — the name describes this lovely corsage or display gladiolus. Striking and expensive-looking orchid-type flower.

**'Velveteen'** — miniature flowered; deep reddish-purple with a marvellous finish like wax-polish.

**'Mahogany'** — medium flowered; deep even 'plain-chocolate' brown. Makes a wonderful centre piece for autumn arrangements.

**'Atom'** — miniature; scarlet butterflies resting on an airy-thin stem.Each floret edged in silver. Delicate and graceful for a one-spike arrangement. In Japan where it was a huge success years ago it was known as 'Beni Kochi' (little butterflies).

For the sensational and unusual large vase or basket display on a suitable occasion there are a few large-flowered cultivars whose inclusion in a mixed colour scheme will cause comment and enquiry:

**'Greenland'** — large flowered, long, thin stems in colour to the tip. Deep leaf-green all the way.

**'Ripple'** — large flowered; mid-blue with small white throat.

**'Dark Night'** — large flowered; black-violet with green throat.

**'Scorpio'** — large flowered; black-red with white lines on each of three bottom petals.

**'Hocus Pocus'** — large flowered; startling colour (not for hospitals) combination: blood-red splashed and striped pure white.

**'Silver Fringe'** — tall, large flowered; opens to the tip. Ruffled strawberry-red flowers edged and striped silver.

**'Globestar'** — large flowered; tall honey-amber with a blue-lilac throat mark.

**'Bright Eyes'** — large flowered; tall bright-yellow with startling red blotch. A ribbon of colour.

**'Fudge'** — large flowered; light-brown with a gold stripe and reddish markings in the throat. Unusual colour combination.

## Fragrants

There is as yet no commercially available fragrant, summer-flowering gladiolus hybrid cultivar, though several people are working on the project. The most fragrant hybrid yet produced is named 'Eau Sauvage', after a French perfume, and is the result of intensive efforts by Igor Adamovic in Czechoslovakia, Mr Adamovic has been involved for many years in producing hybrids between the slightly fragrant gladioli from the USA and the species *G. callianthus* (formerly *Acidanthera murialae bicolor*). His breeding line includes the earliest *Acidanthera* hybrid 'Lucky Star-New Zealand' and a tetraploid version of *G. callianthus*. More specific detail about fragrant hybrids is given in Chapter 8.

## Dragons, Face-ups, Doubles

In very small numbers only there exist certain mutated forms of garden gladioli that are made available from time to time. The original raisers of these unusual forms have long since ceased their work. In the USA old-timers like Kunderd, Doerr and Gosling produced these interesting but most unusual gladioli. Later Mr Oscar Johnson, also of the USA, took the mutations further by selective breeding and produced heavily ruffled, frilled and knuckled forms of floret and named them 'orchid gladioli'. Corms may still be had of some of the more robust of these 'outsiders' or 'off-beat' gladioli such as:

Johnsons — 'Bridesmaid', 'Frizzled Coral Lace', 'Aaron's Exotic Orchid' and 'Wayne Ferris'.

'Doublette' — with the centre filled with petalloid anthers which are all sterile.

Dragons by Gosling with 7 in (15 cm) florets:

'Bird of Paradise' and 'Mauna Loa' — huge laciniated petals on 8 in (19 cm) florets.

Face-ups from various sources in cultivars named 'Red Eye', 'Pink Button', 'Kewpie' and 'Cup-cake'.

# *Chapter 10*  **Classification of Hybrids**

The classification of wild gladioli — the species — has been an ever present problem for botanists and taxonomists for many years. Even now, with nearly 200 species and subspecies fully referenced, there still remain some confusions.

It is obvious that some systematic and scientifically valid method of classifying gladiolus hybrid cultivars is' equally necessary. The present system of classifying hybrid cultivars of gladioli was introduced essentially to cover the various types of cultivars that were used in exhibitions in the various countries where gladiolus exhibitions are staged. The main purpose of the classification system ultimately adopted was to regulate the validity of the entrants in the various size and type classes at shows. It thus follows that there are thousands of cultivars which are unregistered, or not listed in the classification booklets, for the simple reason that the cultivar concerned is not of a type or standard suitable for exhibition. The fact that a cultivar is not registered or classified does not imply that it is not a good or desirable gladiolus cultivar. It simply means that the cultivar is not likely to be seen on the show bench or that the originator did not consider it worth the money and effort needed to go through the mechanics of classification.

There is no international classification for gladiolus cultivars. What does exist and is accepted internationally is the classification system of the NAGC (North American Gladiolus Council). This NAGC Classification is accepted by Great Britain, Canada, Australia, New Zealand and the United States of America. It is a precise and scientific system which classifies gladiolus cultivars according to size of floret, and the predominant colour or colours thereof. The formal gladiolus exhibition so familiar to growers in English-speaking countries is practically unknown in other countries, and only the countries named above use the NAGC Classification to regulate the shows. The NAGC classification is therefore not used in Holland, France, Germany or Belgium, so that the classification does not exist in a metric form. (Even our gladiolus friends in Quebec see the logic of this.)

For the benefit of exhibitors who use the excellent show varieties that the Dutch produce the British Gladiolus Society register and classify any Dutch cultivars that appear on the bench in national or regional shows of the British Gladiolus Society or the Royal Horticultural Society. The NAGC Classification was originally intended for the guidance of exhibitors and show organisers, but over the years it has become internationally accepted by growers, cataloguers and wholesale florists as the 'computer code' by which a coded description of a gladiolus can be instantly conveyed. It is interesting to note that since the NAGC Classification was recognised by the English-speaking countries it has been introduced into the lists of gladiolus cultivars available from the state bulb farms in the Soviet Union and Czechoslovakia, to name but two.

## The NAGC Size/Colour Classification System

The NAGC Classification is indeed a simple three figure system, readily understandable after a few minutes study. The first digit of the code indicates the size of the floret. The floret is measured in inches at the widest point of the widest floret when fully expanded. It is designated as the 'Diameter of Floret', though most exhibitors prefer to call it width since to be pedantic 'diameter' implies a circular floret. For show purposes the size as indicated by the first digit is the classification that governs show entry into the appropriate class. If, for example, a cultivar registered and classified as a 400 size is actually above 400 size on the day of the show it is still acceptable and eligible. To avoid this problem occuring too frequently sizes of new registrations are reviewed after two years of introduction.

The size code key is:

| Description | Size | Width of floret (in) |
|---|---|---|
| Miniature | 100 | Under $2\frac{1}{2}$ |
| Small | 200 | $3\frac{1}{2}$–$3\frac{3}{8}$ |
| Medium | 300 | $3\frac{1}{2}$–$4\frac{3}{8}$ |
| Large | 400 | $4\frac{1}{2}$–$5\frac{3}{8}$ |
| Giant | 500 | Over $5\frac{3}{8}$ |

The second and third digits in the three-digit code relate to colour. The second digit controls and indicates the predominant or basic colour of the floret. If the floret is largely self-coloured, or shades lighter in the same base colour towards the centre, the third digit will be even, i.e. 2, 4, 6, or 8. If there is a significant or conspicuous marking on the floret it will be allocated an odd third digit. An odd third digit will convey that the predominant basic colour is marked by blotch, darts, or marked midribs, flecks, splashes, a contrasting narrow edge to the outer petal perimeter (picotee), or numerous dots of a conspicuously different colour in the throat, (stippling).

## NAGC Colour Code

The Colour Code-key is as follows:

| Colour | Pale | Light | Medium | Deep | Other |
|---|---|---|---|---|---|
| White | 00 | | | | |
| Green | | 02 | 04 | | |
| Yellow | 10 | 12 | 14 | 16 | |
| Orange | 20 | 22 | 24 | 26 | |
| Salmon | 30 | 32 | 34 | 36 | |
| Pink | 40 | 42 | 44 | 46 | |

| | | | | | |
|---|---|---|---|---|---|
| Red | 50 | 52 | 54 | 56 | 58 Black-red |
| Rose | 60 | 62 | 64 | 66 | 68 Black-rose |
| Lavender | 70 | 72 | 74 | 76 | 78 Black-purple |
| Violet | 80 | 82 | 84 | 86 | 88 Black-violet |
| Smoky and other combinations | 90 | 92 | 94 | 96 | 98 Dark brown |

For example, a cultivar classified for the purposes as 478 would have a dark-purple flower with a floret size (large) between 4½ and 5⅜ in across.

The range of colour variations and shades indicated by, for example 70–78, is defined and stipulated in a supplementary document which relates the NAGC Colour Code numbers to the shade numbers as given in the RHS (Royal Horticultural Society, Vincent Square, Westminster, London) Colour Charts.

## Advisory Notes

(a) Greens — light-green and chartreuse: 02, deep grass-green: 04
(b) Yellow — 10 also includes cream or ivory
(c) Orange — 20 also includes buff
(d) Salmon — 36 includes scarlet-orange
(e) Red 50 is not normally used since pale-red is pink or rose
(f) Red 52 — this includes scarlet-reds.
(g) Violet 80 — according to RHS Colour Charts there are no truly blue gladiolus cultivars; such cultivars as approach blue are classified as violet 'dilutions' at 81, 82, 83, 84.
(h) Smokies 90 is tan or beige. Code 91 is the same with markings or shading of grey; 92–96 are brown, tan, mixtures with grey-purple or deep-red, and other combinations; 98–99 are chocolate.

The NAGC Classification was laid down in 1973 and is reviewed from time to time in conjunction with the British Gladious Society, who supply up-dates of the RHS Colour Charts and the reports of the BGS Classification Committee. The NAGC maintains the registration of cultivars and authenticates the size/colour code of new registrations. The Registrar is:

S.N. Fisher
11345 Moreno Avenue
Lakeside
CA 92040
California, USA

Any hybridist wishing to introduce into commerce either personally or through a specialist catalogue is advised to contact the registrar of NAGC in advance as names of cultivars are occasionally duplicated and give rise to confusion.

The British Gladious Society uniquely introduced an addition to the above code system whereby the prefix letter (P) is given to authentic *primulinus* hybrids of show standard. This system only applies in UK since Britain is the only country with classes in their show schedules for the primulinus hybrid. Moves are also being made in USA for a prefix (E) to be approved for registered cultivars of non-exhibition character, i.e. doubles, laciniated, deep ruffled, orchid-flowered etc., to be known collectively as 'exotic'.

## Seedlings for Exhibition

Seedlings not yet in commercial quantity and not registered or classified are eligible to be exhibited in shows organised by the NAGC or the BGS. In the event of an unclassified seedling being exhibited in a size-limited class (i.e. for 500/400, 300, 200/100 classes), the judges have the right to measure the floret, as presented, at the time of judging. The seedling will then be placed in the size-class appropriate to the actual size of the floret measured across the widest part of the largest floret. If subsequently, particularly after trial growing in other areas or other conditions, the cultivar consistently and regularly produces florets of a size different from the original measurement, the cultivar, on request and by agreement of the NAGC Classification, may be re-classified accordingly.

# *Chapter 11*  **The Gladiolus Species**

It is believed that there are approximately 200 different species in the genus *Gladiolus*, distributed throughout the temperate and subtropical areas of the world. They are a fascinating group of plants whose descendants are now probably the most popular and widely grown garden and cut flower in commercial cultivation. A remarkable fact about the modern garden hybrid gladiolus cultivar is that such a diversity of colour, form of flower and size of plant could have been produced from so small a number of the many species available. The entire range of modern gladioli may have had as few as eight parent species.

Professional botanists have made concentrated efforts in the past 20 years to complete a comprehensive review of the information on, and classification of the genus, and have cleared up many of the confusions that arose in the past in the naming of variants and subspecies. Some species with several synonyms have been unified, and certain other allied species, formerly regarded as being of genera other than *Gladiolus* have been re-classified within the genus. This revisionary work has been based mainly on the species of southern and central African origins, so that at present we are still with a somewhat confused list of species from Europe, Asia and north or eastern Africa. Fortunately the southern and central African species account for more than 150 of the 200 plus species already classified, so that perhaps one could truly say that the genus *Gladiolus* is as well documented now as it will ever be.

The two classic books on the genus *Gladiolus* are: *The Winter-Growing Gladioli of South Africa*, by G.R. Delpierre and N.M. Duplessis, published in Cape Town in 1973 by Tafelburg; *Gladiolus: A Revision of the South African Species*, by G.L. Lewis, A.A. Obermeyer and T.T. Barnard, published in Cape Town in 1972 by Purnell.

We do not wish to compete with the botanists, nor to summarise or duplicate any of this valuable taxonomic work. It is most instructive and helpful for those gardeners who have an interest in the species of *Gladiolus* to have a good look at the two books referred to since they between them contain what are almost certainly the best photographs and water colour illustrations of South African *Gladiolus* species in natural habitats. We will attempt to deal, from a gardener's and an enthusiast's point of view, with those species still obtainable that can be used as garden subjects under average garden conditions, or have in the past been used in the hybridisation of the garden cultivars.

To assist those who have grown or may in the future grow the species, we intend to quote extensively from the cultural notes of enthusiastic amateurs and the records of botanical institutions. In particular the records of the late Mr W. Howarth of Accrington, Lancashire, England, will be quoted. His comprehensive notes on his well-tried and successful cultivation methods and his meticulously kept flowering records and descriptions make most interesting reading.

There is no particular botanical significance in the fact that we have decided to divide up the species into four groups on a geographical basis. These groups will overlap in places to some extent, and some of the more common species will occur in two or even more groups. A map (Fig 1 page 9) is included by way of clarification. The groups are:

### The Eurasian Group
These are hardy, spring-flowering species indigenous in Europe, Near and Middle East, and Mediterranean North Africa. (This area includes Israel, Lebanon, Turkey and the Atlas Mountains).

### The East African Group
This group comprises species that are only half hardy and summer flowering, found in Ethiopia, Kenya, Somalia and Madagascar (Malagasy Republic). In this group we include species previously regarded as a distinct genus, the *Acidanthera*, but now classified as *Gladiolus callianthus.*

### The Natalensis Group
This is the group that has played a major role in the hybridisation leading to the modern gladiolus. These species are half hardy and summer flowering and produce the largest plants in the genus. Mainly endemic in central and southern Africa, the species, subspecies and variants in this group were for many years the subject of much controversy as to nomenclature.

### The South African Cape Species
This is by far the largest group, consisting mainly of tender winter-flowering species of slender and delicate proportions. From this group came the inter-species hybrids that were the forerunners of the *Nanus* gladioli. Found preponderantly in the Republic of South Africa, in a triangular area bounded by Springbok (OFS), Cape Town and Port Elizabeth, these are the species that inspired the hybridisation work of Dean Herbert and his contemporaries in the early nineteenth century.

## Cultivation of the Species

It is not possible to lay down any hard-and-fast rules on the cultivation of species that would ensure success. Any cultural advice we offer can be modified, adapted or even ignored in the light of personal knowledge or experience. We give guidelines on cultivation based on methods that have been successfully applied by the authors and others over the years. If there is a single helpful rule it is 'Find out the conditions of the natural habitat and reproduce them as closely as possible'. We define 'conditions' in this context as soil type, temperature, humidity and length of daylight.

In general, the soil required for the great majority of species is poor, light and slightly acid. A well-drained growing medium is essential for all but the marshland species of gladioli. Soilless composts rich in peat should be avoided. The perferred mixture is ¼ vermiculite (perlite) and ¼ silver sand and ½ John Innes Compost No. 2 passed through a ⅛in (30mm)-mesh sieve. Animal manure, particularly fowl manure should never be used. The nitrogen content of the soil must be kept low, and the use of chemical fertiliser at a minimum. An occasional liquid feed of Liquinure or Phostrogen at half the standard dosage is helpful when the flower stems start to develop. Do not, however, use fertiliser or liquid feed on seedlings. Free-draining soil composts as described above are the ideal media for deep pots in which to grow species. The roots of species gladioli require well-aerated soil and an adequate oxygen supply to enable the plant to produce flower stems. The roots must never be allowed to dry out during the growing period.

Despite being small, the corms of species gladioli need to be planted quite deeply so that the moisture level at the roots is never too low. A depth of 2½in (6cm) or 3¼in (8cm) for the larger (*natalensis*) corms is about right. When species gladioli are grown in large pots or tubs a 1in (2.5cm)-mulch of half perlite is useful at flowering time. Not only does this help to support the slender flower stems and keep them upright, but it keeps the soil cool and moist.

During the period of rapid growth keep a check on the pH level of the soil and try to maintain the pH in the range 6.5–7.1. Wild gladioli will be reluctant to flower if the soil is too rich in nitrogen or even slightly alkaline. Instead of flowers the plants will produce extra corms and leaves if the soil conditions are not as suggested. These plants have a very low nitrogen requirement, and an excess of nitrogen, especially as ammoniacal nitrogen ($NH_3$), will result in long, thin leaves and any flowers produced will be sparse and spindly or possibly mis-shapen. Spindly plants resulting from unbalanced nutrition will produce corms of poor storage quality, though this can be corrected if the imbalance is recognised and detected in time. A liquid feed every third day with Phostrogen (formula 22K, 7P, 4N) will supply the high potash level necessary without altering the pH.

The spindly effect caused by excess nitrogen also lowers the plant's resistance to fungal and bacterial diseases. The cell walls become thinner as the cells enlarge, due to excess nitrogen or over-watering, until eventually the cell walls will burst at points of stress. At this stage the leaves will bend and break leaving an ingress for the spores of fungus and bacteria. More species plants are killed by over-feeding or over-watering than die of neglect. Remember, no one feeds the ones in the wild and they bloom their heads off!

Species gladioli are extremely sensitive to the intensity of light and the length of daylight. Careful control of light and water are important factors in raising species from seed and inducing flowers to form. This aspect is of particular importance when dealing with the winter-flowering species from the Cape. Most of this group of species naturally flower during the South African winter (April, May,

June) when the daylight length is about 8 to 9 hours. At whatever time of year the Cape species are planted they tend to flower only in the short-day periods in Europe, that is from late October through to February and March. To simulate the natural habitat of these species it is recommended that as soon as the flower stems and buds appear the daylight length is adjusted by switching on a fluorescent strip light to elongate the 'day'. It works equally if the light is extended in the morning but for convenience we switch on at evening dusk — no doubt present-day greenhouse owners can rig up a timer.

There is no problem of this kind with species of the other three groups as they will flower during early summer months. The Eurasians flower in spring and need no glasshouse protection or light adjustment. In their natural habitat most of the Southern African gladioli grow through a layer of grass, prostrate hardy perennials and low shrubs. The type and density of the ground cover varies according to the soil condition and location, but the cover can often reach 20 in (48 cm). The ground cover appears to be beneficial, one might say essential, to the well-being of the species gladioli; its presence assists the life cycle of the plant in three ways. First, if the ground cover is dense at ground level it keeps the roots cool and prevents excessive drying out. If the associated plants include stiff-stemmed grasses they act as a support to keep the flower stems and the leaves upright. Their presence also helps to lengthen the flower stem, enabling the flowers to bloom at a level where they are in the sun and visible to pollinating insects. Grassy ground cover also provides protection for the seeds which fall to the ground when ripe — safely away from seed-eating finches — and provides the cool, moist conditions suited to germination. Many of the species do not produce flowers in the first season, and the grass protects the tiny seedling leaves from the extreme heat since the sun's rays do not penetrate the dehydrated grasses.

If gladioli species are to be grown in tubs, large pots or in the greenhouse borders, it is often advantageous to leave any adventitious annual grass that has arrived as seed, but do not allow perennial or couch grass to remain. A few light tufts of grass will more nearly duplicate the natural habitat and will help keep the soil cool and moist. Seed, however, must be weeded of all unwanted plants and shaded if the temperature is 80°F (25°C) or above, and the sun is bright and directly impinging on the seedlings.

All but the Eurasian species can be grown in pots; in fact if you have only a few corms that is by far the preferred method. This location ensures complete control of growing conditions and insect attack as the pots can be easily moved outdoors or in, from sun to shade as the circumstances warrant. Pots of plants suffering from insect attack or fungal damage can be readily isolated for the appropriate treatment to be applied. Plants in pots are also much more conveniently handled should you wish to do hybridisation or complete self-pollination and seed collection from designated species. Cormlets too are more readily contained in pot-grown specimens, and can be more readily handled whilst the plant is in a dry condition in the greenhouse despite inclement weather outdoors.

It should be borne in mind that in the pots the soil temperature will generally be higher than in the open ground for any given exposure time to the sun. Black plastic containers or black polythene growbags should never be used in greenhouses for growing species gladioli as they too readily absorb heat from the sun and quickly dry out. Peat-based composts once dried are very difficult to restore to the correct moisture level. Earthenware or clay pots are eminently preferable as they are porous, more robust and less likely to be blown over when the plants are tall. Sinking the pots to half their depth in the garden or greenhouse border is a good way to avoid drying out or overheating should high temperatures be encountered.

A further advantage accorded by pot culture is that the pots can be used as safe storage during the plants' dormancy period. This is particularly useful in dealing with first season corms grown from seed. As soon as the plants show signs of dormancy, or when seed has been collected in the case of mature plants, stop watering, cut out the flower stem and store the dried-out pot on its side in a dry, frost free place. It is immaterial whether it is kept in light or in darkness, but it is very important that the location should have low humidity and a temperature range from 10°C to 25°C (50–80F).

Corms and cormlets stored dry can also be lifted and kept in brown paper bags in a dry, frost free cupboard. Alternatively, they can equally safely be left in the pot until it is time for the growth cycle to recommence. The dry leaves can be detached quite easily from a dormant corm and this obviates the need to take the corms out of the pot. Before storing dormant corms in pots make sure the label is correct and safely attached to the pot. As a safeguard, paint a code number on the side of the pot and record it in your cultural notes.

Species gladioli may be propagated or multipled either sexually or vegetatively. Sexual propagation is by seed. The seed is obtained by successful pollination, the technique for which is exactly as for the hybridisation of garden gladiolus cultivars (see Chapter 7). In order to obtain true seed of a species it is necessary to isolate the plant as soon as the flower buds appear. This is to avoid cross-pollination by any other species that may be flowering at the same time. Any one pot must contain the same species, subspecies or variant, and the pollination should be artificially induced to ensure good setting of seed. Pollen can be transferred from the anthers of the floret to the stigma of another floret using a student's paint brush or a cotton wool swab. If sufficient florets are open together it is better to use the pollen off one floret on the stigma of a different floret. It is seldom that the pollen and stigma of the same floret are ripe simultaneously (Mother Nature's anti-incest policy).

The seeds of all species are best sown in the autumn in the northern hemisphere, preferably in the same soil conditions as the parent species. Some seeds may germinate within a month if the light, temperature and humidity are right, but germination can also be long delayed or even irregular over several months. Never discard any pot or seed before a full year has elapsed since the date of sowing. If seed has been produced by self-pollination within a true species, or cross-pollinated floret by floret in a number of plants of the same species, the resultant seedlings will be identical to the parent species. This is in fact the definitive test for a true species. Should however pollen from a different species or hybrid have been introduced, the result will be a group of seedlings (hybrids) of varied characteristics. The seedlings may show some characteristics of either or both parents but will seldom be identical to either parent. The only certain way to obtain identical plants either of species or hybrids is to make the propagation vegetatively, that is by using the divisions or offsets from the parent corm, or the tiny cormlets that appear in clusters round the bottom of the new mother corm.

Depending upon the size of the offset or division this can be treated as a corm for cultivation purposes. Cormlets are however best treated as seed, with the added step of soaking the cormlets for 24 hours in tepid water to which Benlate (Benomyl) has been added, prior to sowing. Cormlets require to be planted slightly deeper than seed and the maximum depth for the Cape species is 1 in (2.5 cm).

Certain of the species referred to later have been shown to benefit from some modification of the general guidelines on cultivation as given above. These additional comments will be included in the notes and descriptions of the particular species concerned.

## The Gladiolus Species Group by Group

### Eurasian group

The first group of species that we shall examine is best designated as 'The Eurasian species', though this is not a truly accurate description. In this group we place the species that occur in the more northerly areas, mainly European, but also including Asia and North (Mediterranean) Africa.

The predominant characteristic of this group is colour. Of the eleven species only one is not some shade of purple — and that is a white sport of one that is purple. This group contains the only species of gladioli that can be said to be in any way hardy, in that they survive under ice and snow during sub-zero temperature winters in the mountains. Gladioli species from this group have been depicted in art forms and mentioned in literature as far back as Roman and Ancient Greek times. It is believed by some that they are the original 'lilies of the field' mentioned in the Bible. Although the species of this group have been known for several hundred years and have been grown as garden subjects for at least 400 ys, they have contributed little if anything in the way of inter-species hybrids to the make-up of garden gladioli. Many people would say that as garden subjects this group is least favoured and, but for their ability to survive outdoors undisturbed through many European winters, would have ceased to be cultivated years ago. As this condemnation or criticism is well founded and expresses the view of the majority of gladiolus enthusiasts, we shall make no attempt to place these species in order of merit but list them alphabetically.

This list of eleven species does not exhaust the known numbers of the Eurasian gladioli since the USSR claims 30 or more different

*Figure 17 A typical Eurasian species,* G. byzantinus

species scattered through the many republics, and recent Turkish papers mention 38. We intend to concentrate our attention on those we have grown or seen, and know are available for cultivation and study.

**EU-1**

*Gladiolus atro-violaceus*

This species first appeared in botanical records in 1842. Previously it was designated *G. turkestanica* (a designation still retained in USSR), indicating the area where it was first discovered. The area of distribution is widespread and the habitats varied. It is not a rare species, and where it does occur it is well established and plentiful. In Syria, Lebanon, the Jordan Valley and other parts of the Near East this species grows on the edges of cultivated lands or as a weed in fallow or abandoned crop fields. In Turkey, Iran, Afghanistan and the Central Asian republics of the USSR — Armenia, Uzbekistan, Azerbaijan and Turkestan — it is a plant of the mountains, growing on the banks of small rivers and streams or in scrub land by roadsides. It is probably the *Gladiolus* species of the highest altitude and lowest temperature conditions.

As the Latin implies, the colour of the flower is purple-violet, and from 7 to 12 florets are carried on a thin stem which may be from 12 to 25 in (30 to 60 cm) tall. The leaves are flat and very clearly ribbed. It flowers in May or June in lower altitudes and June, July in the mountains. The flowers have elongated lip-petals with feint white lines, and have a partially hooded top petal. Corms are high crowned with heavy husks and are extremely durable. Cormlets develop on short stolons and though few in numbers are quite large. Cormlets will not germinate unless they have been de-husked and are planted the correct way up. Both corms and cormlets are best planted deeply outdoors in September or October. The shoots will emerge in February irrespective of weather conditions. Seed is scarce but if available this is best planted in pots and left outdoors through the winter. Leaves may appear as early as January but frost damage does not appear to affect flowering ability. This species is of interest mainly because of its ability to resist frost and to produce virtually indestructible corms. It is not a suitable subject for pot culture; the flowers are unimposing and short lived.

**EU-2**

*Gladiolus byzantinus*

This species has long been known and was written about in the herbals of the sixteenth century. It is a recorded wild flower in the natural history books of most West European countries, being very common in south east Europe, the Levant and North Africa where it flowers profusely in late May or June. *G. byzantinus* is a garden subject in old-established gardens where clumps of up to a hundred corms have been known to survive undisturbed at least 80 years. The earliest written record of its being grown in Britain is 1768. There is also a white mutation sport designated *G. byzantinus alba*, which, on the rare occasions when it produces seeds, gives a high proportion of the normal purple species in its seedlings.

The plant is 26–36 in tall (60–90 cm); the colour is a medium to dark purple in the upper petals with the lower lip and fall petals having distinct white median lines. The leaves are narrow, ribbed and rounded at the tips, and the first leaf is distinctively 'sword-like'. The stem is frequently branched in mature plants, the main stem having 11–13 florets and the secondaries 5 or 7. The larger flower florets are 2–3 in (5–7 cm) across.

The *byzantinus* is a very sparse seed setter and this characteristic has given rise to a widely held view that it is probably a triploid hybrid of *G. communis* × *G. illyricus*. This opinion is strengthened by the fact that it has been crossed under controlled scientific conditions with *G. illyricus* to give seedling hybrids differing very little from the parent. This species is quite hardly anywhere in Europe and once planted can be left undisturbed for years. It seldom divides corms but produces many cormlets, some as small spawn with others larger and on stolons. The cormlets have a very thick net-like husk which protects the cormlet down to minus 15°C. The plant grows best if planted deep in late September or October. *G. byzantinus* has been recorded in the wild in Britain but such finds are regarded as being chance garden escapes. Corms are commercially available in most suppliers' autumn catologues.

**EU-3**

*Gladiolus communis*

This is the gladiolus of the cornfields mentioned in early Greek and Roman writings, and is depicted in Chinese drawings. It was one of the original species recorded by Linnaeus senior in 1753. The *communis* is widely distributed throughout the Near East, western Asia, USSR and parts of eastern Europe. It is similar to and frequently confused with *G. illyricus*, though when the two are side by side the differences are significant.

Variable in height, dependent upon the habitat, this species is usually 13–25 in (30–60 cm) tall though in grassland or cereal fields it can attain 39 in (100 cm). The plant has few leaves and a pliable stem carrying only 4–8 florets in a single stem, invariably unbranched. The florets are smallish 2 in (5 cm) across maximum, in a pale lilac-purple colour. The classical white medium lines are a feature of the lip and fall petals. There are also white and rose-lilac variants. The normal habitat is grassland or low scrub in poor soil. Flowering time is May to June. The corms are small and flat with thin husks and the production of very tiny cormlets is prodigious. It readily sets seed and probably hybridises with similar species in the wild.

Despite the size of the corm it is best to plant this species fairly deeply 3–4 in (7–10 cm)), and leave undisturbed. If the clumps of corms are covered with a peat or bracken mulch in the winter months (Oct.–Mar.), this little species is hardy down to about minus 10°C. Gritty, well-drained soil is essential as this subject hates wet situations. Seed is best sown in pots in late summer and over-wintered frost free. Corms, however, can be planted in the open in September or October. Protect from slugs or snails in early spring, for if the leading shoot is chewed off there will be no flowers, only leaves.

**EU-4**

*Gladiolus halophilus*

This small and unimpressive species is of interest only because it grows in very wet soil and can even tolerate slight salinity or alkalinity. Its normal habitat is in water meadows or marsh flats. It was first recorded in 1853 in Greece, but it is also known to exist in quite substantial numbers in Turkey, Iran and several republics of the USSR in the Caucasus, Caspian and Georgia areas. Three British botanical expeditions have located and recorded big stands of this species in Turkey — Admiral Furze, Collingwood-Ingram and Anthony Hamilton. Hamilton was guided to the same spot as his two predecessors and found the flowers there exactly as described, at 4,500 ft (1,500 m) above sea level, and flowering in early June.

This species is synonymous with *G. palustris*, which is really saying in Latin what *halophilus* says in Greek. The description of the plant is identical: height 20–30 in (50–75 cm), with 3 to 9 buds. It has very spindly growth with few reed-like leaves. A single rush-like stem carries 3 to 5 small florets slightly upward facing to a height of about 9 in (22 cm). The florets are rose-wine coloured (RHS 74/a) and unmarked. It is spring flowering (May–June) in its native habitat but quite ephemeral. Although, given comparable conditions in cultivation, the corms and seed germinate readily this little species is very difficult to flower. It has been described as 'a challenge rather than an acquisition'. The cormlets are few but quite large in relation to the small conical corms. The corm has a durable double husk which presumably protects it from water and frost damage.

**EU-5**

*Gladiolus illyricus*

*Illyricus* was fist recognised as a distinct species about 1845, before which date there was considerable confusion about it and *byzantinus*. It is widely distributed throughout the Mediterranean area, central Europe, USSR, and it has been recorded wild in France, Italy and Britain.

This is a smaller plant than *byzantinus*, normally 12–20 in (30–50 cm) tall. It has fewer buds, 3–6, is seldom branched and the florets are smaller 1½–2 in (4-5 cm) across. The colour is purple but of a deeper shade than *byzantinus*, the white median lines on lip and fall petals are less distinct and the floret is more tubular. It flowers as late as June or July.

There are several variants of *illyricus*, the most common being colour variations, usually of lighter shades than the normal *illyricus*.

The corms are conical with durable husks, which should never be removed for planting. The large cormlets are usually produced on stolons and require soaking before planting. Plant deeply 4–6 in (10–15 cm) in September/October and leave outdoors undisturbed.

This is not a very fertile species seed being rather rare. Many botanists believe that *illyricus* and *byzantinus* are variants of the same species; and that *byzantinus* as raised and supplied commercially is in fact an *illyricus* hybrid cultivar. However, both are of easy culture and within their colour limitations are interesting garden subjects.

**EU-6**

*Gladiolus imbricatus*

This species was another of those recorded by Linnaeus in 1753 when he first established the genus. Apart from the shape of the floret this is a miniature version of *byzantinus*. Its distribution covers the same area, that is eastern Europe, USSR and parts of Near Asia. It has also been recorded in Italy and Yugoslavia.

A rather smallish plant — only 24 in (60 cm) high — it carries up to 4 florets on a rather stiff stem. The leaves are wide, ribbed and rounded at the tips. The flower stem emerges quite suddenly from the leaf sheath. It flowers rather late for a Eurasian species, normally June or July. The colour is the usually Eurasian indeterminate shade of reddish-purple, the most accurate description is to be found in RHS Colour Chart 78/A. Its habitat is at rather higher levels than *byzantinus* stands, having been found at nearly 5,000 ft (1,500 m) above sea level. *Imbricatus* enjoys poor, well-drained soil and seems to be quite hardy at altitude in European winters.

Cultural requirements are much the same as for *byzantinus*, it is definitely a species for the outdoor border where it will grow with and through lighter-leaved short annual or perennial flowers. Leave undisturbed once established. To increase stock lift and divide about September/October just as dormancy approaches, and re-plant promptly in a new site retaining as much earth as possible. This species has roots all year round even whilst dormant.

**EU-7**

*Gladiolus italicus*

This species is recognised as and recorded as a separate species, but many taxonomists believe it to be so close to *G. segetum* as to be regarded as a subspecies or variant of the more common *G. segetum.*

At first glance one would expect *italicus* to be endemic in Italy but this is not so, for although it is fairly common in the southern Mediterranean it is not particularly well established in Italy. Its distribution is widespread over the southern Mediterranean area, USSR, the Balkans and the Near and Middle East. The plant's habits closely resemble those given in descriptions of *G. segetum* (first officially recorded in 1804). The main difference between *segetum* and *italicus* (if in fact they are not synonymous) is the habitat; *segetum* is a weed of arable land, especially of grain and grass-type cereals crops, whereas *italicus* is much more diverse, being equally happy in untended, rough land of poor soil.

It is a tall, spindly plant with a few small or narrow leaves and a pliable stem reaching 20–30 in (50–80 cm). The simple stem carries up to 10 florets, 6 being the more usual. The florets are placed in a distinctly alternating formation. The colour is light rose-purple with white lines in the throat, on the wing and lip (lower) petals. It is normally in flower in its natural habitat in May or June. In common with most Eurasian species it is reasonably frost resistant in its dormant state despite the diminutive flat corms. It readily sets seed when cross-pollinated with *segetum*, though when grown in pots it is recorded as being 'fairly sterile' to self-pollination.

**EU-8**
**EU-9**

*Gladiolus palustris*

Most taxonomists who have been involved intimately with the classification of the *Gladiolus* species seem to favour the view that *G. Palustris* is the correctly described and recorded nnyyies which Soviet and Turkish botanists call *halophilus*. Both are marsh growers with adaptations specifically suited to their environment, such as the cylindrical leaves, the rush-like growth and the heavily protected root/corm system to prevent water damage.

I have grown both, and apart from the fact that it can grow in conditions which would destroy most other Eurasian species it has very little to commend it. The clever trick that *palustris* has of growing in salt water is not sufficient recommendation for its use as a garden subject. It might be more acceptable if it was some other colour than the Eurasian red-purple.

**Gladiolus segetum**

The botanical description given for *italicus* covers also *segetum*. This species is believed to be the 'Corne Flagge' spoken of and written about by the old herbalists of the eighteenth century, when it was reputed to have curative properties. The corms were either dried or ground up and made into a sort of gruel, and were sold in London street markets.

*Segetum* is readily raised from seed, which is quite large and without the usual 'wings', and if soaked overnight will germinate quite quickly. This is one of the few Eurasian species that readily responds to culture in pots. A flowering-size corm can be as small as ¹/₂ in (1 cm) diameter.

**EU-10**

**Gladiolus tardus**

This is the only known gladiolus species that has a natural habitat in Britain, and there are two locations in the British Isles where it can be seen. The drifts of these flowers in the New Forest about mid-July are probably the most photographed wild flowers in existence, and of course, it is now an endangered and protected species.

This British gladiolus wildling was recorded and registered in 1938 by Worsley under the *tardus* description and was given a specific registration. Controversy rages every July amongst botantists as to whether the flower is entitled to be distinguished with a registration since many of them believe it to be a local variant of *G. illyricus* or a well-established chance hybrid seedling from the garden *byzantinus*.

The plant is obviously winter hardy in its location in light forest conditions and has survived for at least 50 years. In the heaviest frosts the corms deep down at 5–6 in (12–15 cm) are protected by pine needles, leaves and bracken rubble. The corms are of quite high density, distinctly conical and covered by a tough, waxy and net-like complex of husks. The cormlets are similarly protected, and being quite large and carried on 2–4 in (5–10 cm) long stolons, do not suffer from over-crowding when they germinate. Botanical analysis of *tardus* indicates that it has a chromosome number $n=90$, which might explain the paucity of seed, and the low viability of such seed as is produced by self-pollination. A year-round study of these plants in the New Forest habitat reveals that increase is almost exclusively by cormlet, very few seedlings having ever been found at the site. A factor in this lack of seedlings is the peculiarity of the forest habitat, which is not a normal gladiolus *pied-à-terre*.

The edge-of-forest location is not a good one for seed production and distribution of Eurasian-type gladiolus. In order that the seed pods may ripen to maturity warm sun is needed. This species flowers in July and in the normal course of events the seed pods would begin ripening in early September. In the New Forest in September a gladiolus species only 2–3 ft (60–90 cm) tall would be in serious competition to reach the sun's rays, and is fighting against tall bracken fronds and spikes of foxglove (*Digitalis*), ragwort (*Senecio*) or willow herb (*Epilobium*), all of which abound at the forest edge. Early September in Hampshire is noted for misty evenings and night temperature-drop, so that it is quite likely that the seed pods would rot due to lack of sun.

The gladiolus cannot distribute its seeds unless the pod is very dry. The mechanism by which the seed pod opens is triggered by the uneven drying out of three sections of the seed case. As the seed case dries it twists exposing the seed, which in the case of *tardus* are winged to catch the wind. Unless the wings are dry and exposed to the wind they do not detach from the seed pod. The paucity of seed plus the adverse conditions for distribution probably account for the non-appearance of seedlings. It is unlikely that any viable seed ever reaches the ground. The late flowering of its species and its unsuitable habitat could account for the lack of increase.

The issue of whether *G. tardus* is a true species or a variant of *G. illyricus/byzantinus* can only be established by chromosome count and cell matching, or by reproduction from viable seed collected at source. In form and colour *tardus* strongly resembles other Eurasian species, differing mainly in size of flower and lateness of flowering. An interesting botanical challenge awaits any enthusiast who could raise enough seedlings from the New Forest plants to verify its authenticity. Somewhere there must be a young and patriotic enthusiast who could legitimise the New Forest gladiolus and give it a proud title *Gladiolus britannicus* for posterity.

## East Africa Group

The second group to be considered and reviewed are neither botanically nor geographically cohesive. They form a small group of gladiolus species some of which are very well known as garden subjects and attempted parents of garden cultivars. The better-known ones — the former *Acidanthera* genus — have only comparatively recently been included in the genus *Gladiolus*.

For convenience we shall call it the *callianthus* group for the simple and logical reason that the majority of them are the re-classi-fied *Acidanthera* now called *Gladiolus callianthus*.

**EA-1**  *Gladiolus aequinoctalis* — formerly *Acidanthera*. Habitats are south, south-west and west Africa. This species is now quite rare in its former habitat but stocks are maintained in some botanical gardens in South Africa where it grows quite well. This is a very difficult species to grow and flower, and the main interest in growing it is for its quite distinctive and strong fragrance. It is a plant that can reach 39 in (1 m) in height. The thin, rather wand-like, drooping stem carries 5 to 7 florets, star-shaped on a long, tubular neck. The colour is cream with maroon and red markings on the three lower petals. This species was discovered by Baker and registered by him in 1877. Dean Herbert is said to have found it sterile. This is very likely true since it is a triploid (3n=45). A lover of wet, freshwater areas the leaves and corms have developed similar characteristics to *G. palustris* i.e. impermeable husk and reed-like foliage.

**EA-2a, 2b, 2c**  Gladiolus bicolour syn. *G. callianthus bicolor*. This is the reclassified *A. bicolor* of earlier days in basic form and two variants:

2a *Acidanthera* (now *Gladiolus*). *G. callianthus*.
2b *G. callianthus* var. 'Murialae'
This is a larger and improved form of bicolor raised by Kelway of Langport, Somerset, by selection and carefully monitored propagation.

2c  *G. callianthus* var. 'Zwanerburg'.

This was formerly *A. tubergenii*, a self-pollinated seedling from *A. bicolor* raised by the Dutch firm of van Turbergen in Zwanenburg. It differs from 'Murialae' in the floret size — it is slightly larger, and the heavy markings in the three lower petals are purple-red rather than black-maroon. *G. callianthus* and its variants all appear to be triploid (3n=45), but there are some stocks in the hands of hybridists that are tetraploid (4n=60). This *tetra-callianthus* frequently produces good yields of viable seed. *G. callianthus* in one or other of its forms is readily available as corms of flowering size for planting in early May. It is very susceptible to frost damage and also flowers late, thus seed setting is seldom successful outdoors.

All the variants need support at flowering time as they an often exceed 5 ft (144 cm). Large corms, peeled and soaked in water overnight will germinate more readily and flower earlier. A word or two of warning: *callianthus* is damaged by copper-based fungicides (Bordeaux mixture or cuprammonium carbonate).

**EA-3**  *Gladiolus gueinzii* (formerly *A. brevicollis*) Baker.

This is believed to be synonymous with *G. brevifolius* which figured in *Curtis's Botanical Magazine* in 1804 and was later described as *brevicollis* by Klatt in 1882. This species is much wider distributed than *callianthus*, being common in South Africa also. It is vaguely fragrant and has white flowers with mauve and yellow markings on the lip petals. There are no records of its being hybridised even though it has been a cultivated subject under one name or another since 1804.

**EA-3-4-5**  The three former *Acidanthera* which also have changed to *Gladiolus* have little merit as garden subjects. They are not fragrant and the rather insipid, greenish-white flowers have little interest. Like most of the former *Acidanthera* they are triploids.

**EA-3**  *Gladiolus divinis* Marais 1973: Ethiopia, Somalia Eritrea. A species of dry scrubland.

**EA-4**  *Gladiolus amoena* Marais 1973: Guinea, west Africa coast.

**EA-5**  *Gladiolus ukbanensis* Marais 1973: west Africa coast.

The only other species that can be included in this group is

**EA-6**  *Gladiolus garnierii.*

This species is included here more by geography than botany as it is endemic to the Malagasy Republic (Madagascar). Botanically it is closer to the *natalensis* group than any other, and culturally it performs best when treated as a summer-flowering *natalensis*. It was first discovered and recorded by Klatt in 1872 and later as *G. ignescent* by Baker in 1876. The behaviour of this species is conditioned by climatic conditions and variations in Malagasy. On the oceanward side of the island where rainfall is regular and sufficient it is a plant some 25–40 in tall (60–100 cm) with 3 to 12 florets well spaced on the stem. The colour is deep-yellow to orange. In the wet area it flowers May to June in habitats at about 4,000 ft (1,500 m) above sea level. On the drier, landward side the same species grows at lower altitudes on poor, sandy soil. In a very dry season it does not even break dormancy and in the absence of rain can stay dormant for two or more seasons. If there is some rain the plant will produce 1 or 3 florets on a 20 in (50 cm)-stem, set seed and die off in about 60 days. Under drought conditions the small flowers have distinct mauve purple marks on the yellow flowers and are also veined grey.

The main interest in this species is the ability to stay dormant for long periods at high temperature. The corm health is fantastic, so much so that in Holland a research programme is underway to hybridise *garnierii* with some of the older, small-flowered plain cultivars to produce, hopefully saleable cut-flower gladioli similar to *nanus* in form and able to be left *in situ* over winter in suitable growing areas. Crosses with the old 'Acca Laurentia' and 'Hopman's glory' cultivars show promise. As *garnierii* in its different forms shows a range of different colour pigmentations it is quite likely that hybrids of other colours could result. In Holland and in the author's garden *garnierii* is treated like a primulinus hybrid cultivar and it responds well. It sets seed readily, produces good-size cormlets in numbers and germination of cormlets is good. It tolerates dry conditions well. Not much is known about *garnierii* and there is no commercial supplier.

## The Natalensis Group

We must preface this survey with an apology. We regret that much of the descriptive information will be tinged with nostalgia and may be emotionally distressing to those who grew the *natalensis* gladioli many years ago under their old names. For those who missed out on Latin in their formative educational years we point out that *natalensis* means 'orginating in Natal', and this in itself adds to the confusion since many of the plants we describe in fact come from places other than Natal.

Whilst we are sure the taxonomists are right, they have created some difficulties by changing the names of many beloved old friends. Most of the confusion that occured before the revision of the genus *Gladiolus* was caused by botanists and explorers personalising the species as they discovered them. True botanical binomials are much to be preferred.

A second apology concerns the 'order of appearance' in non-alphabetic sequence. It is more convenient to begin with the master species, *G. natalensis* — the one that started it all in 1756.

This is a widely distributed species with many variants and sub-species. It occurs throughout all but the desert areas of Africa from the eastern Cape Province (RSA), through tropical central Africa to Ethiopia and western Arabia. In one or other of its forms it is now established as a wild flower or garden escape in Europe, the southern states of USA and parts of Australia and New Zealand. Most readers would probably have come across the species as *psittacinus*, so named because at the early stage the flower bud emerges to resemble the beak of a parrot.

The first descriptions of the type species noted that it was tall and summer flowering. Height varied from (3 to 4½ft) (1 to 1.5m), the leaves broad, 5–12, and usually in a fan-shaped pattern with the flower stem emerging from the sheath of the leaves. The florets number from 12 to 25 with 3 to 6 open at a time, the placement being more usually alternate rather than in parallel rows. The flower pattern is distinctive, the uppermost petal is the largest and is fully hooded or partially hooded. The two petals below the hood are wide, pointed and form wings. The three lower petals are about a half to one-third the size of the upper three. Any markings are invariably on the three lower petals and may be median lines, blotches or darts of a different colour. Any throat markings are normally lighter in shade than the main petals.

It is the flower form, petal colours and the throat markings that separate one variant from another. Petals may be mainly yellow, orange, red, rose, green, brown or combinations. Purple or lilac is rarely the main colour in the *natalensis* group. Throat markings may be streaks, mottled or fine lines. Some variants have fine veining of a second and different colour.

The Southern African master species (ecotype) has been given quite a number of subspecies designations over the years as collectors found variants in new locations. Corms of the ecotype are believed to have been brought to Cape Town from Natal in 1823 or 1824. Ecklon saw the first flower in the garden of a Mr Smuts in 1826 and called it *Gladiolus ecklonii* (as was the fashion of the day). The colour of that flower was orange, changing to red-speckled yellow lighter in the throat. That was the flower now regarded as the genuine home-grown *G. natalensis (forma domesticus)*. Dr Ecklon thought the flower was a *Watsonia* and reported that it had a fragrance similar to lilac, a point which has never subsequently been confirmed.

In 1827 Dr Dalen of Rotterdam received some corms of the same species from Port Natal, and as Dr Dalen was at that time the Director of the Rotterdam Botanic Gardens he had them in flower there in 1828. Dr Dalen had illustrations made and captioned the picture as — guess what? — *Gladiolus dalenii.*

In a magazine article in 1830 a Mr Hooker illustrated his contribution with yet another picture of *natalensis* but this time Hooker noted the profile of the buds at opening stage and called it *G. psittacinus*. Between 1830 and 1837 two new colour variants of *psittacinus* arose and were described, one from Mr Hooker and the other from Mr Cooper. These variants were known as var. *hookerii* and var. *cooperii.*

The *hookerii* variant is quite a distinctive progression from the original *G. psittacinus* in that it is a penta-ploid (5n=75). It grows over 6ft (1.75m) tall and its clear-yellow flowers are 4in (10cm) across. The variant flowers very late, taking 190 days from planting to flower. (The author has it flowering under glass in November). It seldom sets seed as there are no suitable specimen pollinators flowering at that time and it appears to be self-sterile. Propagation is no problem as it produces huge cormlets up to ¾ in (2cm) in diameter and they flower first season. There is now a rose-coloured *hookerii* with similar characteristics.

After 1870 the var. *hookerii* was very widely distributed all over the world. It almost ran wild in certain locations propagating so rapidly it became a nuisance. In Western Australia and parts of South Australia it became established as a weed when gardeners who had imported corms as a 'fashion craze' wheeled away barrowloads of corms and cormlets to bury them in the ditches. The var. *cooperii* was less prolific and smaller with a muddy greenish-yellow colour of flower.

Another variant arose about 1876, a maroon and green-speckled type named *dracocephalus.* The descriptive Latin is apt, the flower is narrow and pointed and in early bud stage certain resembles a reptilian head. The smaller and better clear-yellow variant from the Victoria Falls area was introduced in 1902 under the name *G. primulinus* (primrose coloured). From this beauty came the new strain of primulinus hybrids created mainly by British hybridists Kelway and Unwin.

Collingwood-Ingram found and distributed a variant very similar to *G. primulinus* in all aspects except colour. Whereas *primulinus* was a clear primrose-yellow the Ingram type was cloudy buttery orange. It was *primulinus* heavily veined with orange lines which at a distance gave the cloudy effect, hence the Latin *nebulicola.* of *G. nebulicola.*

Dr Lewis found a similar variant in Zambia which was without the veining but of a pure buttercup-yellow. It was fully compatible with other yellow *natalensis* types, producing good cultivars in the lighter shades. Dr Lewis named this *G. xanthus* and it is claimed that this was the parent of the first white primulinus hybrids.

There are other species that could be included in the *natalensis* group in that they are summer-flowering types from the same area. We did say that we would concentrate on those species that were either good garden subjects or had been used in the creation of garden hybrids, and the species we have described qualify on both counts. If they are treated in the same way as summer-flowering, small-flowered cultivars they are rewarding subjects.

All of the species described are best grown 'naturally', that is no de-eyeing and no heavy feeding. The larger *natalensis* species will attain 6 ft (1.75 m) without feeding, but they will require supports.

*Figure 18  A dominant parent of the modern gladiolus cultivar.* G. cardinalis

## The South African Cape Species

The majority of the known and verified gladiolus species are indigenous to southern and south-east Africa, and within this majority are almost 90 that occur in the Republic of South Africa. We will concentrate on those species which have been known to culti-vation outside South Africa for a century or more, and particularly those species that have contributed to the development of the summer-flowering, outdoor garden gladiolus cultivars.

In looking at, and admiring a garden gladiolus today, one must feel compelled to dwell on the thought that its original ancestors were being first cultivated almost 175 years ago. At the same time one must feel a little diminished by the concurrent thought that since 1930 not much has really changed. A most suitable starting point is the group of larger-flowered species found beyond the Cape Province of South Africa, but also in that area, and which provided the basis for the development of the 'grandiflora' cultivars of the present.

This group is indisputably headed by:

**Gladiolus cardinalis**

(The Waterfall Lily, or New Year's Day flower in Afrikaans). Both common names inform us of the habitat and flowering period in South Africa. Seen in its native habitat it is the most striking member of the group with its scarlet-red flowers and contrasting clean white darts on the three lower petals. *Cardinalis* grows in upland areas near waterfalls, wet cliffs or on the banks of streams and is in flower during January. The arching stem carries 5–7 buds, with the flowers actually reversing upwards on the stem. Easy of cultivation, it requires a soil rich in humus and to be kept moist, preferring half shade. It is a robust plant with thick, dark-green leaves. If grown in pots or tubs it is best left undisturbed. The propagation of corms and cormlets is very slow as the new corms produced each year are not detachable. Any disturbance is resented, frequently causing the plant to miss blooming for a season immediately after transplanting. This characteristic is still retained in many of the *nanus* hybrids with *cardinalis* parentage.

**Gladiolus saundersii**

Another parent of modern gladioli, *saundersii* is in many ways similar to *cardinalis*. The stem, however, is erect, the buds 5–7 in number. It is shorter than *cardinalis* and seldom opens more than one floret at a time. The uppermost petal is plain light-red and is recurved rather than hooded. The remaining five petals are almost equal in size, the three lower ones having white median lines and white patches liberally stippled in the same light red as the petals. The corm is flattish globose, and leaves arise fan shape from the upper central part of the corm. Few cormlets are produced from the corm but this species sets seed well by self-pollination. *Saundersii* is one of the very few species where the ripe stigma falls below the pollen-loaded anthers, thus facilitating pollination.

**Gladiolus sempervirens**

The interesting feature of this species is its habit of having little or no dormancy period and being virtually evergreen. The leaves develop in a wide fan of 10 to 12 leaves and the root is, in appearance, intermediate between a corm and a rhizome. It also has a number of perennial fleshy roots that elongate to increase the spread during flowering time, and though the stem is thin the flowers are held erect even in the strong evening winds. It has been collected by waterfalls at 5,000 ft (1,700 m) in stony mountain slopes.

**Gladiolus sericeo-villosus.**

This is one of the largest plants in the group. It regularly grows 6 ft 6 in (2 m) tall. The leaves grow in a fan of 7 to 9 and are very wide. The erect flower stem carries upwards of 40 buds in the placement of a foxglove (*Digitalis*) and the florets are upward facing. The petals are about 2 in (5 cm) long arranged cup-like with two petals much larger than the others. Unlike other members of the group, the flowers are somewhat unattractive in both colour and shape. There are several colour variants but all are in rather dull combinations of pink, lavender or purple-red streaks in a cream background. Several hybridists, including Dean Herbert, used this species in an attempt to increase the bud count of hybrids. Alas, it proved not to be an arithmetic success whether used as pollen or seed parent. There are no statistics recorded, only a sad and regretful comment.

### *Gladiolus oppositiflorus* (Herbert 1837)

This is another of the species which Dean Herbert very successfully introduced into his hybridisation programme. He describes it as 4½ft (1.5m) tall, growing well and forming clumps where the old corms persisted year by year strongly attached to the upper corm. Modern growers have also discovered this and the absence of cormlets, so whilst *oppositiflorus* is not difficult to grow, increase other than by seed is virtually impossible. One is left amazed that people like Herbert grew these plants in Yorkshire 150 years ago, while today with all our additional scientific advantages we struggle to keep them alive.

This robust and very productive plant grows not only tall but has numerous quite large flowers of long-lasting quality. Up to 23 buds are quite common and of these 15 or 17 may be fully open at the same time. Unfortunately, that is if one's interest is in gladioli for exhibition, the placement is not like other species. The large flowers are outward facing and in pairs. The top petal is somewhat hooded and plain coloured whilst the other five are median lined in a darker colour. The basic colour is lilac-pink, but there is also a light-salmon variant. In its favour, however, this species readily sets seed and they germinate freely.

It was Dean Herbert's aim to produce hybrids with more open flowers simultaneously and of longer-lasting quality. The fact that we now have modern cultivars with 25–27 buds, opening 10 in flower, is in no small measure due to the efforts of the hybridists of earlier times who persisted in using high bud count species.

### *Gladiolus cruentis* (1868)

This species was used on the earlier hybrids, such as the *gandavensis*, to improve the colour saturation and to introduce the blood-red factor. *Cruentis* is in the parentage of many of the good large-flowered reds for which Pfitzer and others became famous. The chief asset of this species is the intensity of its blood-red flowers. Though in form the florets are rounded and cup shaped pollen from *cruentis* markedly improves the colour on other paler coloured hybrids. It is a smaller plant than others in its group, being only 15–30in (36–90cm) tall and carrying 3–10 florets. It is robust and can be seen at altitude in mountainous areas up to 6,000ft (2,000m). It is a late-flowering species in South Africa, blooming as winter approaches. It is stated that some of the hybrids sent to Queen Victoria's garden at Osborne House were *cruentis* offspring.

The final part of our review of species covers the smaller and more delicate members of the genus endemic mainly to the south-west corner of the Republic of South Africa and to that part of Africa now known as Namibia.

The species are, in appearance, quite dissimilar to the other groups, being closer to the freesia at first glance. They are more slender, with leaves often rush-like and cylindrical and seldom exceeding 3ft (1m) in height. They flower in the winter in the rainfall areas of Cape Province and all have a distinct dormancy period following their rather short flowering season. This group has the widest variation in colour form and growth patterns and contains many fragrant species. Some are rare and verging on extinction, most are easy of cultivation given the right conditions, set seed readily, produce cormlets and inter-breed without difficulty. We begin with:

*Gladiolus carmineus*

Known as the 'cliff gladiolus' this species was used in the earliest of inter-species hybrids. The flower resembles that of *cardinalis* though the red is less intense, and fewer florets are borne on its short, erect stem (8–12 in) (20–30 cm). The flowers are 2 to 5 on the stem and each about 2 in (4.5 cm) across. The species *carmineus* is found on sea cliffs in damp, shaded areas often in very poor soil. A surprising feature of this species is that the flower stem appears before the leaves, which are long, trailing and widely ribbon-like. Seed, which germinates readily, is available to members of the South African Botanical Society on a regular basis.

*Gladiolus maculatus* Small Brown Afrikaner

This species is more freesia-like than most, the similarity being enhanced by the strong fragrance. In Africa it is autumn flowering and fairly common in its various colour forms. The basic ecotype is heavily speckled brown on yellow, its small tubular flowers are about ¾ in across (2–2.5 cm) carried 2 or 3 per stem. The pink and reddish-brown subspecies occur on sandy plains and are without fragrance. It is best raised from seed, leaving the tiny corms to complete their dormancy in the dry pot.

The fragrance of *maculatus* is not usually transmitted to hybrids by pollination.

*Gladiolus odoratus*

Very similar in appearance and height (12–15 in) (30–40 cm) to *maculatus*, this species, despite its name, is less fragrant than *maculatus*. It has 6–9 florets on the slender, erect stem, the florets being about 1 in (2–3 cm) across. The colour is predominantly brownish-purple, though closer inspection reveals a basic creamy-yellow heavily marked and speckled in purple, brown and maroon. The fragrance is transient, lasting only a few hours when the stigma is ripe. *Odoratus* sets seed readily and it germinates well. It is an early-flowering type rather than winter flowering.

*Gladiolus carinatus* Blue Afrikaner

This is one of the most commonly occurring wild species of *Gladiolus* in the Cape Province and appears in many colour forms. To describe it as blue is really an exaggeration, at best it is yellow with blue-grey edges to the petals and blue-purple lines at the back. There are several colour forms, the colour varying according to the soil conditions. The form most sought after by hybridists is the one from the sandy soil of Cape Flats. It is strongly fragrant and red-purple. The upland form is also fragrant; the colour of this one is yellow with greyish-mauve markings. The size of the plant also varies according to habitat ranging from 6 to 24 in (15–60 cm) in height, flowers 2 to 9, flower size is consistently 1 in (2–3 cm) across.

It is easily grown from seed in sterile, sandy compost using silver sand, for preference.

The species that can lay claim to being blue are the 'Riverdale Bluebell' and the 'Caledon Bluebell', spp. *G. rogersii* and *G. bullatus* respectively. Writing about these two beautiful flowers is akin to composing an obituary. Only once in forty years have I seen a flower of these species and that under cultivaion. Both *bullatus* and *rogersii* carry single, blue-purple bell-like flowers on a thin stem some 12 in (30 cm) tall. The flower is like the single floret of a blue foxglove (*Digitalis*), even including the stippling in the throat. The leaves totally surround the stem in a sheath, and are fine and rush-like. It is impossible to pick the flower without taking all the leaves. This is what happens to 'the bluebells' in habitat. Deprived of leaves and flowers the corm dies, leaving no cormlets, whilst the seedless flowers wither in the living room of some thoughtless vandal. Unless some conservation measures are promptly taken these two beautiful and irreplaceable species will die out.

Both are difficult to cultivate but it can be done from seed, which is occasionally made available from conservation societies in South Africa (see Chapter 1).

**Gladiolus gracilis**

This is a very common wild gladiolus of South Africa and the flowers are vaguely scented at night and during the day. There are various colour forms, one of which is called 'Blue Sandpipe' in Afrikaans. A truer colour description would be blue-grey overlaid lilac. Mature plants carry 4 florets on a 10 in (24 cm) stem. The florets are about 1 in across (2–3 cm), and the petals have a pronounced point at the tips. The slender, rounded leaves and the fine, flexible stem give the whole plant a graceful air, hence the Latin *gracilis* — graceful.

This species was amongst those used for inter-species hybridisation by Dean Herbert and by Professor T.T. Barnard in his 'Purbeck hybrids'; and several fragrant hybrids have been created using *gracilis* × *tristis*.

**Gladiolus liliaceus**
**Large Brown**
**Afrikaner**

This species has not only a striking appearance but also a remarkably sweet fragrance by night or evening. It is likely that the species is pollinated by night-flying insects. The colour of the flower changes perceptibly as evening approaches, and it is the opinion of some botanists that the throat of the floret may emit fluorescence visible to insects.

Although the flowers are tube-like the petals are pointed, recurved and very narrow. From 2 to 5 florets are carried on a slender stem of 20 in (50 cm). The elongated petals give the florets a 2–2½ in span (5–6 cm) which is large for the size of the plant. It is winter flowering (July) and survives in heavy soil on the plains and mountains. It is easy to cultivate in greenhouse, or garden in sheltered areas. *Liliaceus* occurs in a yellow, a russet-red and a chocolate-brown and yellow form in different locations.

The hybrids of *liliaceus* × *tristis* are evening fragrant, confirming that the description 'lily-like' refers to the fragrance, not the flower form.

137

**Gladiolus caryophyllaceus**

This is the species that has the fragrance of carnations and motivated Dean Herbert to try to introduce the fragrance into the gladiolus cultivars of the early nineteenth century. As Dean Herbert was at the same time as he was hybridising gladioli species, also hybridising dianthus pinks, he considered this particular fragrance most appropriate. (See page 21.)

It is a very attractive plant, with a longish stem of light-pink, bell-like flowers which individually may be 2 in (5 cm) across. The arching stem may carry eight or nine florets all but two of which will be open together. The overall height of the plant is often 2 ft (60 cm) in good conditions. It is not difficult to cultivate in cool, well-drained, sandy soil and is readily raised from seed. During dormancy the corms are best left in the pots in a dried-out state. If damp the corms will quickly succumb to rot. The best preventative is flowers of sulphur raked into the top layer of soil. Never use copper fungicides. Benomyl has proved ineffective on dormant corms.

**Gladiolus alatus**
**Turkey Lily**

It is not quite clear why this species is called the Turkey Lily as there is no resemblance to the turkey either in colour or form. It is, however, a striking and colourful little species of only 4–6 in (10–15 cm) height. Despite its lack of height it puts on a great floral display with its 4–7 tightly packed florets one above the other on the stem. The bottom floret is 2 in (5 cm) across with a most intriguing shape and colouration. The top petal is reddish-brown and hooded and the large, spreading wing petals are of the same colour. The three lower petals are narrow and protrude well beyond the hood giving a rude greenish-yellow, tongue out effect.

It grows in the heavy soils of the cultivated plains where several variants can be found including a rare one with a totally lime-green 'tongue'. It is not fragrant, and though it is easily cultivated from seed there is no record of its having contributed to the garden gladiolus cultivar parentage.

**Gladiolus orchidiflorus**
**Green Turkey Lily**

This is a sweetly fragrant and delicate species in the same form as *alatus* and has attracted the attentions of hybridists for many years. With *G. watermeyerii* these two species constitute the most spectacularly colourful pair of dwarf species in existence. *Orchidiflorus* is still fairly common but *watermeyerii* is quite rare.

*Orchidiflorus* grows from 6 to 10 in (15 to 25 cm) high and usually carries three 1 in (2.5 cm) florets on the stem. The three dorsal petals are greenish-brown and the 'tongues' of the other three petals are lime-green marked maroon — an effective and truly orchid-like combination. *Watermeyerii* has the same statistics but the dorsal three petals are spectacularly striped brown on lime-green whilst the tongue is beige and bright golden-orange. Seed of *orchidiflorus* is available more frequently than that of *watermeyerii*, which is extremely scarce.

**Gladiolus tristis**
**Yellow Marsh**
**Afrikaner 'Evening Scent'**

This lovely freesia-like gladiolus species was one of the very first to come to Europe from South Africa. It was registered as early as 1754 by Linnaeus. It was used in the earliest of hybridisations and is the true forerunner of the garden gladiolus. One can say that *tristis* and *cardinalis* are the pillars on which garden gladioli were founded. It occurs in many forms from 20 in to 70 in (50 to 179 cm) tall and in colours from yellow to deeply striped brown on cream. The tall form closes its flowers all day, and most forms are fragrant in the early evening. The florets from 3 to 10 are widely spaced on the stem and are from 1–2 in (2.5–5 cm) across. The tubular florets have all six petals of equal length and tend to face outwards.

*Tristis* readily sets seed and is easy to cultivate. In sheltered temperate areas it can withstand low temperatures during dormancy though it is not truly frost hardy. It has been used extensively in the inter-species hybrids and was the basic parent of many of Dr Barnard's Purbeck hybrids. As a transmitter of evening fragrance it is the No. 1 species.

*Gladiolus angustus*
Painted Lady

This is a species of the uplands where it grows by the side of mountain streams. It is winter flowering, reaching a height of 30 in (70 cm) at flowering time. More usually the flower stem falls over and hangs down. To compensate for this the florets reverse and face upwards. The stem carries up to 10 florets about 2 in (5 cm) across. The wing petals are elongated, narrow and pointed and in a clear cream or lemon-yellow. The three lower petals are blotched with spade-shaped markings of yellow-edged dark brown. This species was used in the early hybrids and it is from this the blotched cultivars originated.

*Gladiolus carneus*
Painted Lady

Not only is this species confusing because it is also called Painted Lady, but because it occurs in so many different forms and variants. The many variants at one time had different names and at least four of these have now been included in *carneus*. This species is also the one that was in earlier times *G. blandus*.

If one were to see the variants side by side it would be excusable to think they were different species:

*carneus* is cream with three yellow and reddish-purple blotches;
*carneus* (*blandus*) is pale-pink with black-purple full blotches;
*carneus* (*albidus*) is without any blotches and is white.

All three forms in their various guises were used to hybridise what are now known as the *nanus* hybrids. All the forms are without fragrance in the wild.

*Gladiolus
floribundus*

This is another species with a wide distribution throughout southern Africa and occurring in several forms and variants. Its plant characteristics are very close to *carneus* except it is essentially upright growing. This species has at least four recognisable variants in which colour and height are the main variants:

Subspecies *millierii* – height 15 in (40 cm) plain white flowers upward facing.
Subspecies *fasciatus* – height 8 in (20 cm) white base heavily marked, blotched and lined red-purple. The 3 or 4 florets are in what is almost a terminal umbel giving a totally unique formation of a creeping gladiolus. This subspecies is under cultivation and

hybridisation in California, where it is being developed as a ground-cover garden gladiolus.

Subspecies *miniatus* — height 24 in (60 cm) larger, plain rose-pink flowers.

Subspecies *rudis* — height 15 in (40 cm) light-pink flowers with three yellow and lilac blotches on the lower petals.

It is quite likely that this species in one or other of its forms was used by the Dutch in the early days to produce the *nanus* and *gandavensis* hybrids. These hybridisation programmes took place before the revision of the classification of the genus *Gladiolus*, and thus may have been recorded as *G. millerii, G. rudis, G. miniatus* etc.

We are left now with about 150 species that are recognised as definitive species about which we have no information as to their use in hybridising. That is for the future historians to consider.

## Allied Species and their Hybrids with Gladiolus

Several of the plants now listed as *Gladiolus* were originally thought to be of other genera, for example:

*Homoglossum*
*Watsonia*
*Antholyza*
*Babiana*
*Acidanthera*
*Ixia*

The very first African *Gladiolus* to be cultivated in Holland was believed to be, and cultivated as a *Watsonia*.

There are two recorded hybrids between *Gladiolus* and other genera:

*Acidanthera bicolor* × *G. hortis* c.v. 'Filigree'
*Homoglossum aureum* × *G. ornatus*

*G. aureus* is the link species with the genus *Homoglossum*.

It would seem that after almost 200 years of the hybridisation of the gladiolus less than 20 of the 200 species are known and recorded as having been used in the creation of our present-day garden gladiolus. In making this observation we must pay tribute to our predecessors for their dedication and persistence. In so doing we hope it may initiate some enthusiasm in those who still have the energy, ability and skill to continue this work. May, one day, Dean Herbert's vision become a reality.

*Figure 19 Gladiolus inter-species hybrid of 1820-1830. (Tristis, carneus and cardinalis parentage)*

**Specialist Societies**

There exists throughout the areas where gladiolus can be grown a number of specialist gladiolus societies, all of whom welcome members with an interest in the flower.

A special welcome is always accorded to those enthusiasts from countries where no specialist gladiolus society exists. Special arrangements are made for overseas members to receive year books and bulletins by air mail where available, subject to an additional amount *over* normal fees to cover postage costs.

In addition to specialist societies there are one or two organisations that issue bulletins or magazines devoted to gladiolus information — such as recently introduced cultivars, show results, and tips on cultivation.

The main specialist societies are:

**The British Gladiolus Society**
Secretaries:
JK & Mrs E Anderson
20 Rock Avenue
Nailsea
Avon BS19 2AJ
England

**Yorkshire Gladiolus Society**
Secretary:
I.M. Sunderland
4 Meadow View
Skelmanthorpe
Huddersfield
West Yorkshire HD8 9ET
England

**Canadian Gladiolus Societies**
**Vancouver Gladiolus Society**
Secretary:
P.Q. Drysdale
3770 Hardy Road
R.R. Number 1
Agassiz
British Columbia VOM 1AO
Canada

**Frazer Valley Gladiolus Society**
Secretary:
Grant Wilson
1274, 129A St
Ocean Park
Surrey
British Columbia V4A 3AD
Canada

**Winnipeg Gladiolus Society**
Secretary:
T.J. Musty
140 Braemer Avenue
Winnipeg, Manitoba R2H 2K7
Canada

**New Zealand Gladiolus Council**
Secretaries:
Mr & Mrs R. Wilcox
13 Ramanui Ave
Hawera, Taranaka
New Zealand

**Hawkes Bay Gladiolus Society**
Secretary:
I.A. Lean,
309B Gascoigne St
Hastings, HB
New Zealand

**South Australian Gladiolus Society**
Secretary:
L.J. Ellis,
60 Mitcham Avenue,
Lower Mitcham 5062
South Australia

**North American Gladiolus Council**
Membership Secretary:
Reinhold Vogt
9338 Manzanita Drive
Sun City, Arizona 85373
USA

**Western Australia Gladiolus, Dahlia and Bulb Society**
Secretary:
Mrs J. Tonaut
42 Kathleen Street
Nth Cotteslow, 6011
Western Australia

**Gladiolus Commercial Growers Division**
Secretary:
Dr R.O. Magie
6501 17th Avenue W. W403
Bradenton
Florida 34205
USA

Membership of the above societies is on the basis of a renewable annual fee.

The All-American Gladiolus Selections issue coloured booklets and leaflets illustrating current and past cultivars still available as 'All-America' corms at regulated prices. Please send postal costs equivalent to $2 US for the information sheets to:

Secretary
Sam Fisher
11345 Morena Avenue
Lakeside, California, 92040
USA

A tremendously informative catalogue in full colour with over 60 full-colour illustrations is the world's best gladiolus buy. For $2 US (or its equivalent) Carl Fischer will send his masterpiece.

Noweta Gardens Gladioli
Prop. C. Fischer
900 Whitewater Avenue
St Charles, Minnesota
USA

The Gladiolus Breeders Association issues three bulletins a year devoted to the hybridisation of both species and cultivated garden hybrid gladioli. The GBA also arranges to hold stocks of seed of many wild South African species. Full details and current membership fees are available from:

Secretary:
F.N. Franks
15 Guildhall Drive
Pinchbeck, Spalding
Lincolnshire, PE11 3RE
England

# *Appendix II*  **Where to See Good Gladioli**

The short answer to the question 'Where can one see good gladioli?' is 'at gladiolus exhibitions'. This is apt and true but does not really answer the question.

There is really nothing for gladioli like the displays *en masse* one sees in Holland and Lincolnshire in England for spring flowers. There are no Gladiolus Rally events like the Tulip Rally in Holland, and no decorated vehicles with gladioli as there are in the spring flower fairs. It is quite true that the major regional and national shows of the British Gladiolus Society are an opportunity to see gladioli hybrids at the peak of perfection – but the cultivars one sees there are limited. Present day gladiolus shows lack the commercial displays of earlier years when suppliers vied with each other to put on show – non-competitively – the widest possible selection of the cultivars they list and supply. The lack of commercial stands at shows is due to a very large extent to the high cost of mounting a display, transporting the flowers and display furniture and, if the flowers are imported, also meeting the stringent plant-health requirements.

If the flowers cannot be economically brought to the public then the public must go to the flowers – and that sadly is the situation today. Let us begin by describing what you are likely to see at major gladiolus shows and where and when they take place.

In Europe one seldom sees exhibition gladioli at shows until the end of July or early August. The first major gladiolus exhibition in Britain is staged at the Royal Horticultural Society Halls in Westminster, London SW1. This is indeed a specialist gladiolus show and is normally scheduled for the second Tuesday in August. The schedule is most comprehensive, affording the opportunity for exhibitors to show gladioli cultivars in all the classified sizes and types, including the primulinus hybrids. The highlight of this RHS gladiolus competition is the major class for the honour of winning the Foremarke Cup. To win the Foremake Cup at Westminster is the ambition of every gladiolus exhibitor of note.

The class calls for 12 specimen stems of any cultivar of any size or colour – though traditionally the entries are invariably in the 400–500 class sizes. The rules permit only one cultivar to be exhibited in duplicate, so that a visitor is sure to see at least eleven cultivars of the highest possible calibre. There are medal awards for the Best Spike in The Show (Champion Spike), usually a large-flowered cultivar, and also a Champion small or miniature cultivar. The exhibits include a wide range of cultivars, and the three spike and individual spike classes are most interesting. The show is at the beginning of the gladiolus season and is the ideal date for the blooms of the primulinus hybrids. In our opinion the season's best primulinus are generally seen at Westminster.

The National Gladiolus Show organised by the British Gladiolus Society is usually about the middle of August, the date varying according to the venue. The BGS is committed to a policy of staging the National Show in a different region of the United Kingdom each year on a rota basis, though technical and transport difficulties have prevented Northern Ireland from providing the venue as yet. The BGS National Show Schedule runs to over 50 classes most years, covering the widest possible range of colours and types of gladiolus cultivars. All classes are well contested in an average season, giving a fine opportunity to both novice competitors and visitors to see what can be done with the flower in the hands of experienced showmen.

The most valuable trophies to win at the National are The Blake Trophy and the Daily Mail Gold Vase. The Blake Trophy is for 6 specimen spikes in the large or giant-size flowered cultivars and is carried away by the proud exhibitor of six spectacular and imposing gladioli approaching 6 ft (173 cm) high. The Gold Vase — presented to the BGS about 1928 by the proprietors of the *Daily Mail* — is now of astronomical value for the gold alone. It is not awarded every year because of the nature of the exhibit required. The requirement is extremely difficult to obtain, calling as it does for three matched stems in full flower of a seedling cultivar raised exclusively by a British Gladiolus Society member in Britain. The criteria for the award are that the seedling must show some advance in development as regards colour, shape or form, such as to enhance the gladiolus as an international flower. Past winners have produced outstanding primulinus hybrids, the first laciniated-petalled flower, the early fragrant gladioli, the first green and brown cultivar and several spectacular exhibition large-flowered cultivars. The co-authors have each had the honour of holding the Gold Vase at different times.

Under the auspices of the British Gladiolus Society regional shows are held in late August to early September. These shows are well contested in the many classes scheduled, most schedules catering for seedlings and primulinus hybrids. At one regional show each year the Gladiolus Breeders' Association, (a BGS-affiliated society) put on special classes for seedlings produced by members not only of the Breeders' Association but also of the BGS. The Southern Regional Show is at Twickenham, the Northern at Leigh, Lancashire, and the Yorkshire is at Harrogate. The Scottish Regional Show is at Ayr. For those who like to see flowers other than gladioli there is also a very fine show at Southport.

Any of the shows held under BGS rules are well worth a visit, and this is probably the only way one can learn about the cultivars and their potential. Importantly, it also gives would-be exhibitors an insight into the best cultivar to grow to produce show-quality flowers for a particular show at a specific date. It is a trifle strange that the Netherlands, so famous for its colourful tulip fields in May, has no comparable display fields of gladioli in August, despite the fact that an estimated 20 million gladioli are raised annually in the bulb-growing area.

One can, in fact, see thousands of gladioli flowers in August in Holland, but not growing in fields. Daily at the Aalsmeer Flower Auction, near Schiphol Airport, thousands of bunches of cut-flower gladoli can be seen from six in the morning to midday. Similarly at the fields, one can see thousands of flower-heads in huge piles by the side of the irrigation ditches. The flowers are cut off with stems and await the visiting inspector.

To protect the quality of the gladiolus corms that ultimately go for export world-wide, the Dutch Government have strict regulations for growers. A registered gladiolus grower can either grow gladioli for the cut-flower market or for corm production. If the grower chooses to grow for the flower auction he must pull up the entire plant at the appropriate stage of flowering, chop off the corm and roots. Later, in the presence of a plant inspector, he must destroy the roots, usually by shredding or pulping. The cut-flower grower has access to bulk certified growing stock at special rates the following year only if he has complied with the bulb-health requirements. Similarly, the corm producer must plant only certified stock of registered cultivars and as soon as the gladioli show enough colour in the buds to verify the name of the cultivar the flower stems must be removed. In the case of new cultivars or stock believed to be mixed he may leave the blooms to mature and then chop them off. Again, the cutting and culling must be done with the inspector's authority and presence.

There are, however, ways of seeing Dutch gladioli. At the sales nursery of Franz Rozen at Vogelanzang, near Lisse, the Rozen family have a large display garden for summer flowers. There the Dutch gladioli raisers may send their new or recently introduced cultivars to be grown in big display beds of up to 200 corms per cultivar. The cultivars are clearly labelled and the originator's name is shown. From the end of July to mid-September a very good show of gladioli, begonias, dahlias, outdoor freesia, lilies etc. are there for the public to see. There is also a very well-stocked and superbly organised sales section where orders can be placed for bulbs, corms, or tubers to be delivered at the appropriate time.

Gladioli can also be seen at Haarlem at the Bulb Growers' meetings from the end of July until mid-September. The gladiolus competition is on alternate Tuesdays, covering five Tuesdays over the period. Raisers of gladioli may send as many cultivars as they wish to the meeting and display them in basket arrangements of between 20 and 30 stems per cultivar. A floral committee judges the flowers, and evaluates them on a points basis. Each cultivar may be displayed twice in the six-week period and the average points are taken for the final assessment. At the end of the six-week period the points are calculated and a Silver Beaker is awarded to the best large-flowered cultivar, and a Silver Trophy to the best small-flowered. The results of all the assessments and details of the flowers concerned are printed in the bulb growers' trade magazine. Interested visitors — bona fide of course — are made welcome and attractive translator/guides are usually present; but go before 2 pm or it will be all over.

If you go to Aalsmeer Flower Auction don't expect to see lovely show-standard flowers. Everything at Aalsmeer is produced for the cut-flower local and export trade, particularly to London and German cities, so that all the gladioli are in tight bud in wrapped bundles of ten, or boxes of tightly packed stems in 50s or 100s.

## North America

The vast area covered by the North American Gladiolus Council — from Canada and Alaska to the Mexican border — gives great scope for seeing gladioli *en masse* for most of the year. Naturally the best quality flowers and the more recently introduced cultivars are to be seen at the shows which start in early June and end in mid-September.

Even after the shows have ended it is still possible to see gladioli in thousands growing in the open at, for example, California and Florida. At Oceanside, California, the Frazee organisation grow and export flowers literally by the million and always arrange to have a large crop of cut-flower gladioli running into hundreds of thousands in 40 or 50 different cultivars on their extensive sand-dune holdings by the Pacific Ocean. From late November until the end of January the Frazee organisation are sending truckloads of gladioli to the mid-west and north-eastern States. By arrangement with the proprietor or his sons, Mr Ed Frazee will welcome genuine gladiolus fans to the flower fields and warehouse, where in the depth of a northern winter one can forget the ice and cold and see not only gladioli but freesias, orchids, lilies, proteas, calla, gardenia, carnation and dozens of other exotic species.

In the Bradenton Sarasota area of Florida the Manatee Fruit Company have huge tracts of citrus-fruit groves bearing in the months of November to January magnificent oranges, lemons, grapefruit etc. and the blossom for the following year's fruit. Between the rows of citrus fruit trees, in lines of 500–600 metres long, the MFC have hundreds of rows of commercial gladioli for the cut-flower trade in the northern and central states of the USA. In the mild and favoured climes of Florida it is indeed a thrill to see the flowers of our summer in full glory being loaded on the trucks. Visitors are welcome to the flower areas to buy flowers at summer prices.

Shows in the USA and Canada usually cater for all types and colours of gladioli and any of the major shows are worth a visit, The 'Recent Introduction' and seedling classes are of special interest since the best hybridists tend to send the best of their creations to local shows. One may see a potential Show Champion a year earlier at a local show.

Approximate show dates in USA and Canada are listed below:

Mid-June — California, South Carolina
Early July — Sacramento (Col), Missouri, Illinois
Late July — Nebraska, Illinois, Colorado, Iowa
Early August — Indiana, New York State, Missouri, British Columbia (Canada), Michigan
Mid-August — Massachusetts, Pennsylvania, Idaho, Ohio, Wisconsin, Saskatchewan (Canada)
Late-August — Manitoba (Canada), Idaho, Rhode Island,

Washington State, Utah, Quebec (Canada), Ontario (Canada), Minnesota, Maine
September — Eastern New York

There are late shows in the southern states of California, Florida, Arizona, Texas, and New Mexico in late September, but here the gladiolus exhibits are part of the floral art section rather than exhibition gladioli. It is, however, very interesting for those gladiolus lovers who use the flower of the gladiolus in arrangements. Many novel arrangement ideas are in evidence at these late shows, particularly in relation to compatible foliage and other flowers. A camera is a must for these shows, but to get full true-colour value use fast film and blue (north light) flash bulbs.

# *Appendix III* **Suppliers of Cultivars and Species**

The following are specialist suppliers of named gladiolus cultivars and gladiolus species. Many of the gladiolus species are available only as seed. The list is not intended to be fully comprehensive but has been compiled with the aim of advising readers where the expert and experienced suppliers may be found. Particular reference is made to established suppliers who specialise in the various types of gladioli. Where suppliers are also growers they frequently trade only wholesale and these firms are prefixed (W).

## Summer-flowering Gladioli

### United Kingdom

Walter Blom and Son, Leavesden, Herts, WD2 7BN England
Show and garden summer-flowering cultivars. *G. nanus*, *G. byzantinus* and allied genera i.e. *Schizostylus*, *Watsonia*, *Crocosmia* etc.

De Jaeger and Sons, The Nurseries, Marden, Kent, TN12 9B England
Collections of summer-flowering gladioli cultivars for exhibiting, landscaping and cut flowers. Butterfly, small flowered and primulinus included. The autumn catalogue includes several named cultivars of *G. nanus* as well as *G. byzantinus* and *G. byzantinus alba*. On request they will supply from Holland the small-flowered, winter-flowering gladioli suitable for tubs on patios, in conservatories or frost-free greenhouses.

(W) Jacques Amand, Clamp Hill, Stanmore, Middx, HA7 3JS England
Retail with discounts for 100 corms per cultivar. Summer-flowering gladioli for garden and floral art including Butterflies, primulinus hybrids and miniatures. *G. nanus* type (available for spring flowering) supplied in October/November, some *nanus* and *colvilleii* types are retarded and made available in February/March for July flowering. *G. byzantinus* and *G. byzantinus alba* are available in November for immediate planting.

Sunglads (Ian Sunderland), 4 Meadow View, Skelmanthorpe, Huddersfield, West Yorkshire, HD8 9ET, England
A recently established mail-order organisation personally directed by an enthusiastic exhibitor with long experience of show management and schedules. The catalogue lists only Dutch cultivars, mostly show cultivars. The Sunglad policy will hopefully go a long way towards solving the vexed problem of 'substitute cultivars', for Mr Sunderland intends to publish his catalogue later than most, and only when he has firm assurance that listed cultivars are available. The intention is

to comply at all times with customers' requirements as to size of flower and colour classification.

(W) Peter Nyssen Ltd, Railway Road, Urmston, Manchester M31 1XW, England
A huge stock of Dutch cultivars of all types including the Visser primulinus hybrids and *tubergeni, colvilleii/nanus* hybrids. For the benefit of semi-commercial growers and exhibitors this firm will usually supply 100 corms of a cultivar at the wholesale rate.

Broadleigh Gardens, Bishops Hill, Taunton, Somerset, England
The supplier targets more at the florist, flower arrangers or specialist gardener, in that they have an extensive list of species suitable for garden or greenhouse as well as the species hybrids, *nanus, colvilleii* and similar. The list also includes many closely related species such as *Watsonia, Antholyza, Schizostylus, Ixia, Montbretia, Crocosmia* etc.

(W) J. Parker, Dutch Bulbs, 452 Chester Road, Old Trafford, Manchester, England
A comprehensive list of the regular, commercial summer-flowering gladioli namely of the older-established cultivars. A few of the more recent introductions are occasionally on offer. This firm concentrates on Dutch cultivars in ·bulk qualities, excellent value for garden centres and commercial growers.

Showglads (A.J. Smith) 3 Baddeley Drive, Wigston, Leicester LE8 1BF, England
Top-size and high-quality corms of selected exhibition cultivars from Holland, Australia, USA and Canada. The proprietor is himself a successful exhibitor at national level. His free advisory service is extremely helpful in giving flowering-time expectations cultivar by cultivar. Some overseas cultivars are acclimatised, having been grown in Europe for at least one season. It is well worth the additional pence to have reliable corms at the right planting time and of the ideal size.

Bloomfield Glads (J.A. Webster), 28 Bloomfield Gardens, Arbroath, Tayside DD11 3LJ, Scotland
Specialising in summer-flowering exhibition gladioli. Extensive list of established show cultivars from Holland, Canada, USA and Australia, with miniatures (100 and 200 size) included. There is a good selection of floral art cultivars too. Corms are trialled in Scotland as to suitability to perform in northern areas, and the Australian cultivar corms are fully acclimatised. This proprietor is also a Gold Medal Exhibitor.

Rainbow Glads (J.D. Marston), 5 Woodlands End, Lepton, Huddersfield HD8 OHY, England
Specialist in exhibition summer-flowering gladioli including miniatures and primulinus hybrids. Listings in this catalogue are based on show records at the major UK shows over the past five years. Recent cultivars are introduced only after evaluation in trial gardens in

Yorkshire. A Gold Medal trade exhibitor in his own right, he also produces an excellent subscription magazine, *The Rainbow Connection*.

## The Netherlands

(W) N.V. Konynenburg and Mark (K & M), Offenweg 42, Noordwijk, Binnen, NL
Dutch-grown cultivars of K & M origin are mainly offered but a few cultivars of other raisers are also listed. Excellent value for commercial growers or garden centres. Still the best supplier of 'Green Woodpecker'.

(W) Preyde Bros, Molenvaart, 4, Anna Paulowna, Zeeland, NL
Specialists in large-flowered commercial gladioli of their own-raised cultivars. Some of their giant-flowered (500 size) cultivars can be grown to huge show-winning proportions. For this purpose the Preyde Bros will supply Jumbo (16cm circum.) bulbs on request.

(W) J.P. Snoek, Directorenweg 41A Ems-Nop 8307 PC, NL
Probably has the finest exhibition cultivars ever to come out of Holland, and as these cultivars became better known (and more widely distributed) they will be winning championships for years to come.

(W) Van Tubergen, 86 Koninginnenweg, Haarlem, NL
Specialists in species hybrids, *G. nanus, G. colvilleii* and some of the historic *tubergenii* hybrids which were originated by this family firm. Retail orders can be supplied via the UK branch office in spring. A specialist firm for the gladiolus connoisseur — address: V. Tubergen UK Ltd, PO Box 16, Diss, Norfolk IP22 3AA, England.

## Canada

(W) Leonard Butt, Huttonville, Ontario, LOJ 1BO Canada
Lists over 200 cultivars of worldwide fame. Introduced the first ruffled miniature gladiolus over 40 years ago and has the best selection of these cultivars of any supplier. His catalogue includes the All-American selections of the current and five previous years and the Top Ten show cultivars of Canada and USA. Recently he imported stock of the best 20 Australian show gladioli from Ray Podger. These 'Aussies' are grown on in Ontario and British Columbia to acclimatise to northern hemisphere conditions before sale. This stock represents the best way of getting hands on the big Aussies, that is big in bud count — up to 31 buds in some cultivars.

## United States of America

(W) Noweta Gardens (Carl Fischer), 900 Whitewater Avenue, St Charles, Minnesota 55972, USA
The most sensational coloured catalogue of gladioli is published each year by Carl Fischer. Until quite recently the photos in Kodakcolor were by Carl, the floral-art arrangements and the script still are. Noweta Gardens has supplied more All-America Selection gladioli than all other hybridists together.

For floral art, landscaping and commercial florists, the Noweta gladioli are outstanding — some make good exhibition spikes too. The results of 50-year hybridising have made Noweta world famous, with their cultivars being grown in the Soviet Union, the countries of Western and Eastern Europe and South America. Strong on deep ruffling and blotches in unusual colour combinations, Noweta glads please all but the ardent exhibitor.

(W) Selected Glads Inc. (AJ Nagel), 18389 Michigan Ave, Three Rivers, Michigan 49093, USA
Al Nagel is the general manager for this firm, who grow 3 million gladioli every year. Selected glads grow on all the exclusive and patented stock of the All-America Selection. This collection of North America cultivars undoubtedly represents the best all-round gladioli that exist today. At least two of the eight or so cultivars selected each year end up in the Top Ten show list. Small-size (8–10 cm) (dia. $^3/_4$ in) corms of the All-American cultivars can be bought quite reasonably, and if grown in good conditions can make top-size 'dollar a bulb' show glads for the following year's shows. Do you know an entrepreneur with 5 hectares (11 acres) of sandy soil in the New Europe of 1991?

Pleasant Valley Glads, (G. and R. Adams), 163 Senator Avenue, Agawam, Mass. 01001, USA, PO Box 494
This father and son organisation has been in the gladiolus business for many years and has been responsible for the introduction of many fine gladioli. Their list is quite extensive, covering all types and sizes of gladioli. Patient and always willing to please, they will supply even two corms per cultivar. The emphasis of the Adams catalogue is to give new hybridists a chance to capture the USA market if their flowers are good enough. For the exhibitor the Pleasant Valley list includes the USA Top Ten Show Collection and the 10 best-selling All-America Selection — the key to the trophy cabinet, they say!

Summerville Glads, RD1 Glassboro, New Jersey 08028, USA
A list for the connoisseur and exhibitor. Alex Summerville seems to snap up all the worthwhile USA seedlings that have show potential. He also supplies 'breeder' glads for the hybridist. His recent introductions each year include some fantastic new glads at equally fantastic prices — up to $5 each. There are many of the more recent show winners available from this firm at realistic prices and the bulb health and quality is normally excellent.

**Australia**
R. and K. Podger, PO Box 87, Camperdown, Victoria 3260, Australia
This young couple who took over the former raising business of Errey Brothers are now probably the only suppliers left in Australia who are willing and able to supply the northern hemisphere. Quarantine and plant-health regulations are very strict in Australia and few growers will undertake exports, especially as this involves cool-storing bulbs for nearly 6 months to get the right dormancy time. The Podgers list many Dutch, Canadian, USA and New Zealand cultivars. However, the reader's interest is more likely to be centred

on the huge 29–33 bud, giant exhibition cultivars the Aussies do so well.

When choosing from the Podger's list remember that Australian judging standards are quite different from those of Britain or North America. The Australians have Formal, Informal and Intermediate classes. The Informal Champion in Adelaide even on a good day wouldn't get a 'Very Highly Commended' at Southport. Some really beautiful glads are available in Australia, but try to see a photo of the cultivar before you pay $2 a corm plus air freight charges.

## Gladiolus Species and Species Hybrids

The availability of true gladiolus species corms has much diminished in recent years. Increasing sensitivity on plant health and stricter control on importation of declining or endangered species has added to the problem of obtaining true and healthy stock from overseas. Political difficulties in Southern Africa have made the collecting of plant material from the Veldt and mountains an extremely hazardous operation. There is also a shortage of trained botanists who can reliably identify and collect the required species. Very few of the South African nurseries that dealt in native South African wild flowers still operate, so what little material is available is now from private gardens or the various gardeners of the South African botanical societies throughout the RSA.

It is very likely therefore that any species material from the RSA would be in the form of seed. Species seed is made available to members (overseas only) of the Botanical Society of South Africa and is collected from the gladiolus species growing in the Society's gardens. The address is:

National Botanic Gardens of South Africa
Kirstenbosch, Newlands, Cape Province RSA

A limited amount of corms of the European and Eurasian species are still readily available, mainly grown in specialist gardens in Holland and Belgium. The growers in Holland who maintain stocks of *Montbretia, Crocosmia, Homoglossum, Ixia* etc., grow *G. nanus, G. colvilleii, G. byzantinus, G. callianthus* (*Acidanthera*), *Schizostylus* and similar allied species. The following list of suppliers of species material is given in good faith and was valid at the date of publication:

National Botanic Gardens (address above)
Species seed, mainly of the Cape winter-flowering types, is available to overseas members of the Botanic Society. A seed list is sent twice a year to eligible members, and there is a list devoted entirely to bulbous subjects in which the available *Gladiolus* spp. are given. The list normally includes about 40 to 50 species and subspecies of *Gladiolus* and allied genera (i.e. *Watsonia, Homoglossum*). The number of packets of seed each member may have varies from year to year, as do the available species.

Van Tubergen (UK) Ltd, PO Box 16, Diss, Norfolk IP22 4AA, England

Usually list *G. callianthus* (*Acidanthera bicolor murialae*) named cultivars *nanus* type. *G. byzantinus* and *G. byzantinus alba*, as dry corms. The *G. byzantinus* forms and most *G. nanus* are available in November, the *G. callianthus in spring.*

Chiltern Seeds, Bortree Stile, Ulverston, Cumbria LA12 7PB, England Seed only available by post. Cape and other South African species *G. orchidiflorus, G. scullyii, G. cardinalis, G. tristis, G. alatus*; mixed South African species with up to 10 different species in the mixture. Eurasian species: *G. segetum, G. illyricus, G. imbricatus.*

Avon Seed, 29 Nore Road, Bristol BS20 9HN, England Corms and seeds of species and subspecies from the wild — mainly South African. Eurasian species available as corms in October/ November. Corms of *G. colvilleii, ramosus, nanus* are also listed as species and are available October.

## Czechoslavkia
Ing. Igor Adamovic, Suhvezdna 10, 82102 Bratislava, CSSR One of the foremost hybridists in Eastern Europe. His activities came to light in 1976 when a seedling he brought from Czechoslovakia won the National seedling trophy at the British Gladiolus Society's Golden Jubilee Show at Westminster. Adamovic now has a range of his own exotic and fragrant cultivars, including some *G. callianthus* hybrids. Some of these cultivars were displayed at Stoke Garden festival in 1986 and one of his exotics, 'Bambino', won three Blue Ribbons in the USA in 1987. For the collector of art nouveau in gladioli these Czech cultivars are a must — expensive but uniquely different.

This list of suppliers in no way implies that other suppliers not listed are inferior or unreliable. The author has over 40 years experience of growing and buying gladioli of all types. This list synthesises and condenses that experience in a shortened version, giving an ample range of sources that should enable the reader to locate the gladioli of his or her choice. Together with the List of Cultivars and the List of Species it should be possible to find enough information to secure the chosen stock.

Garden Centres may occasionally have good prices, but one often finds that these are quickly snapped up by 'those who know' leaving 'DDD' quality (dubious, defective or doubtful) to be rummaged through less gently than desirable. The only way to be sure you get the cultivars you want, true to name and from good healthy stock, is to go to a specialist firm. The firms quoted in this list have proved reliable and invariably deal in cultivars that have been evaluated and proved in test gardens supervised by experienced gladiolus growers.

Should you order from any of the firms in the list, it is helpful to give in advance a few cultivars you are prepared to accept as substitutes or alternatives should your chosen cultivars not be available. This will avoid you having to accept a substitute that you do not want or already have. Alternatively, you could indicate that in the event of one of your choices being not available, they increase the numbers of the available cultivars. Remember, when buying gladioli corms, it takes just as much time and costs much the same in fertiliser, sprays etc. to grow a poor gladiolus as it does a worthwhile and rewarding bloom.

**Pests and Diseases**

## General Advice

Gladioli of whatever type are no more susceptible to disease or prone to insect attack than any other popular garden subject. Provided reasonable care is given and simple precautions taken, it is quite easy to grow good, healthy gladioli from corm planting to lifting without any problems whatsoever. Not only can you save your stock from year to year, but with the application of good lifting, curing and storing techniques you could increase your stock.

It would be no help at all to the potential gladiolus enthusiast to develop and discuss all the pest and disease possibilities that might affect his plants. That would be an immense task and quite unnecessary. Not only would it take up valuable space that could be better used on other matters, but it might frighten away a possible champion exhibitor. The analogy that comes to mind is the advice of a midwife to the new mother — she always tells the mum what to do, and it is quite unknown for the midwife to present the new parents with a medical dictionary of all the possible ailments and diseases the baby might suffer!

The basic essentials to good gladiolus culture are simple enough but must be strictly adhered to in order to retain good healthy stock. The first essential is to be sure to buy new stock from a reputable source. Most established suppliers obtain their stock from registered growers whose entire crop is twice tested during the growing period and finally certificated as free from pests and diseases (including lethal virus) before it is shipped or exported. Never introduce in to your garden any gladioli corms that you may buy cheaply at street markets or car boot sales. This is a source of much diseased and virus-ridden stock, especially if the corms are bought late in the season (end of May to end of June). It is not unknown for unscrupulous traders to deal in corms that are uncertificated or have failed to meet the critical health requirements needed to obtain the Certificate of Phytosanitary Health. Reputable traders will burn unsold corms at the end of May as a safeguard against the spread of diseased corms.

The second essential is to plant the corms correctly in suitable conditions and at the right time, and to give the corms a good insecticide and fungicide dip before planting if you have any qualms at all about the health of the corms.

Thirdly, be ruthless in destroying any diseased or obviously virus-affected corms or plants at any stage of growth, irrespective of the original cost or sentimental value of the corm involved. Maintain good garden hygiene at all times and watch carefully over your stock during the storage period. Remove any diseased, doubtful or dubious bulbs from storage and keep the suspects segregated until planting time. Then you may be able to decide whether to plant them or not. If left in contact with other corms during storage a bad corm can quickly affect all the others in the same tray or bag.

Much damage has been done and many good gardeners frightened off growing gladioli by extravagant stories of the difficulties and problems of pests and diseases in growing them. Against that scenario we have many hundreds of growers who have kept stocks of their favourite cultivar for up to 20 consecutive years and maintained or increased the numbers of corms over the period.

The problems that affect gladioli in culture can be roughly classified as pests, (insects and similar), fungus diseases, and virus and deficiency symptoms.

## Insect Pests

The insect pests that may from time to time attack gladioli can be readily classified as those whose detrimental activities take place in the soil and involve attacks on root or corm, and those which attack particular parts of this plant during the growth period. In the former group are nematode and eelworms, cutworms, wireworms and the larvae of ground beetles.

### Soil-borne Insect pests

Wireworm and leatherjackets eat the new roots as they begin to develop, and though their activity is not fatal to the gladiolus, the plant is retarded and recovers only slowly from major attack. These two pests and their near relations are usually associated with grassland. Particular care should be taken to deal with the possibility of their presence, in particular if you are growing gladioli on land newly dug from grass field or lawn area.

*Nematodes* There is just a possibility that you could encounter eelworm or nematode worm. Although there is no known specific eelworm or nematode that attacks gladioli exclusively, the potato eelworm and nematodes from chrysanthemum and phlox can attack gladioli too. The obvious answer to potato eelworm is not to plant gladioli where potatoes or tomatoes grew the previous season, if you can do so. If not, then you may have to resort to soil fumigation or sterilisation as recommended (see Appendix V).

*Cutworms, beetle grubs* Most other under-soil pests are the larvae (grubs) of beetles of various kinds. These grubs activate rapidly in early spring and have become voracious feeders on roots and bulbs by the time the corms are due for planting. A dusting of insecticide in the planting hole will see them off.

*Keel slugs* In very wet, slow-draining ground where fertility is maintained by liberal application of manure or a fully organic compost such as from a compost heap, keel slugs are a problem. This particular slug is seldom seen above ground but if you turn up a small clod of earth the keel slug will be seen. It is about $\frac{1}{2}$ to $\frac{3}{4}$in (1 to 2 cm) long, of a light-cream to white colour. Its rasping jaws can quickly damage the main shoot of the gladiolus allowing rot fungus to penetrate. As this slug does not eat slug pellets it has to be killed by liquid slug killer (aluminium sulphate).

## Insect Pests on Plants

The major threat to gladioli by way of insect pests is undoubtedly thrips (*Taeniothrips simplex*). Adult insects are about $1/12$in (2mm) long and are highly mobile. If viewed with a hand lens the insect is seen to be rather elongated and banded white and black, the minute clear wings being scarcely visible. The real villain of the piece is the larvae or 'wingless thrip'.

The eggs of the thrips are laid inside the leaf sheaths or bud scales on the upper part of the plant. The young thrips are sap suckers and as they appear in hundreds the damage they do is irreparable. Signs are silver streaking of the foliage accompanied by brown tips and shrivelling. On flower stems the bud scales are spotted white or brown with flowers similarly marked, often distorted and failing to open. Since the pest is living within the leaf sheath or flower bud, spraying is not much use. Systemic insecticide is the only effective cure and this must be applied in very good time as a preventative, since although the insecticide will kill the thrips it will not repair the leaf and flower damage.

*Aphids* (*blackfly and greenfly*) Occasionally gladioli plants in full growth may collect colonies of aphids about the end of June. This occurs more often in gardens where other flowers are grown and the aphids have migrated from roses, chrysanthemums, calendula etc., or from broad beans (blackfly). If spotted in good time before the aphids have had a chance to enter the flower buds the treatment is simple and effective. For those who have no qualms I find the most effective way is to put soft cloth in hand, dip it in liquid derris and gently remove and liquidate the colonies in a careful upwards sweep to the tip of the plant.

*Caterpillars* There is no more effective way of dealing with caterpillars than using the same weapon and gently crushing the invader. Caterpillar attack is invariably the work of a solitary larva so the need to spray the entire patch seldom arises. Caterpillars are easiest to detect in the early evening. If, however, you are the rare victim of multi-caterpillar attack a spray of rotenone or derris is effective.

*Whitefly* Whitefly do not normally attack gladioli grown outdoors. If they are seen they should be sprayed immediately with Dimethoate (systemic insecticide), and treatment repeated at 4-day intervals until clear.

## Fungus Diseases

There are four main fungus diseases of gladioli, and though it is rare for anyone to have corms or plants affected by all four in any season, it is well to be prepared should any one of these occur.

*Fusarium yellows* Probably the best known of the fungus diseases is *Fusarium oxysporum var. gladiolus*. There is no visible way of knowing whether a corm of gladiolus has *Fusarium* spores within it or not. *Fusarium* lies dormant and gives no external signs of its presence, and the only way to know if the stock of a gladiolus cultivar is infected is to see the plant in full growth just before flowering time and again just before the lifting of the corms. This is the basis of the Plant Health Certification system. Inspectors from the Plant Health department, or its equivalent in the growing country, visit registered growers at the appropriate time and subject

**Chemical Agents in use in Gladiolus Culture**

## Warning

All chemicals used as pesticides, insecticides, fungicides and soil fumigants must be handled with the greatest care. *Read the label carefully before use*, and make quite certain that the chemical agent you intend to use is safe for the particular plants you wish to treat and that the climatic conditions are suitable. When treating the *Gladiolus* species or *Gladiolus nanus* cultivars it is safer to decrease the concentration slightly as these types of gladioli are much more sensitive than their robust heavyweight brothers.

## Soil Treatments

Soil-borne diseases and pests can be eliminated by cleaning up the soil in which the gladioli are to be planted. The label will state the time that must elapse between treatment and planting. Adhere strictly to the advised time as failure to do so may cause yellowing of the leaves on the first shoots; this in turn retards the growth because of the chlorophyll deficiency.

Never grow gladioli on the same land for two consecutive seasons. If, however, it is not possible or practical to avoid it, then you may have to resort to soil fumigation.

*Soil fumigants and drenches*

| Chemical | Trade name | Pest or disease eliminated | Use |
|---|---|---|---|
| Metam-Sodium | Vapam | Cutworm, wireworm, fungus | |
| Dazomet | Mylone | | |
| Methyl bromide | — | Nematodes plus above | Soil fumigant or drench |
| Dichloran | Botran | Fungus | |
| Disystan | Systemic Granules Jeyes Fluid | Inspect and sap-sucking pests, especially thrips | |
| Phenol (carbolic) | Lysol, San Izal | Multipurpose | |

*Soaking dips for pre-planting corm treatment*

Phenol

| | | | |
|---|---|---|---|
| Benomyl | Benlate | | Soaking dips |
| Thiram | Arasan | Multipurpose | for pre- |
| Dichloran | Botram | Fungus only | planting corm |
| Captan | Orthicide | | treatment |

*Pests during growth. aphids and whitefly*

| | | | |
|---|---|---|---|
| Dimethoate | Systemic | Aphids and whitefly (blackfly, greenfly, rose aphids) | |
| Derris-rotenone | | Many (search the label) | During growing season |
| Permethrin | | | |
| Malathion | | | |
| Pirimiphos | | | |
| HCH, derris, thiram | Hexyl | | |
| HCH, BCH | Gammexane | | |

*Caterpillars and weevils, chewing insects*

| | | | |
|---|---|---|---|
| Derris-rotenone | Many | Caterpillars, weevils, chafer, grubs, froghoppers, etc. (chewing insects) | During growing season |
| HCH | Gammexane | | |
| Permethrin | Many | | |
| CH, derris, thiram | Hexyl | | |
| Dimethoate | Various trade names include 'systemic' | Thrips | Repeat spray at 4–7 day intervals |
| Parathion | | | |
| Permethrin | | | |

*Fungus disease during growing period*

| | | | |
|---|---|---|---|
| Benomyl | Benlate | Botrytis and similar | |
| Dichloran | Botram | | During growing season |
| Mancozeb | Dithane 945 | | |
| Thiram | Arasan | | |
| HCH, derris, thiram | Hexyl | | |

*Sprays for corms after lifting but before storage*

| | | | |
|---|---|---|---|
| Captan | Orthocide | Fungus *Stromatina* | Sprays for corms after lifting but before storage |
| Thiram | Arasan | *Sclerotinia* | |
| Phenol | Lysol, Jeyes Fluid, San Izal | *Botrytis* | |

*Dormant corms in storage*

| | | |
|---|---|---|
| Bordeaux mixture | Fungus and dormant insect eggs or overwintering aphids and immature thrips | Dormant corms in storage |
| Copper fungicide | | |
| BCH Dust | | |
| Malathion dust | | |
| CH, derris, thiram | Hexyl Dual Purpose | |

*N.B.* Benlate is not considered a suitable safeguard on dormant corms as this product acts mainly systemically during growth period.
To prevent the growth of blue or green moulds or mildew on dormant corms use flowers of sulphur or naphthalene flakes.

# Bibliography

## General reading on Gladiolus culture

*The Gladiolus Annual*, the year book of the British Gladiolus Society — first issue 1926. Issued free to members, but can be purchased.
*The North American Gladiolus Council Quarterly Bulletin*, (four issues a year). Issued free to members, but can be purchased.
*The World of Gladiolus*, Koenig and Crowley (eds), Edgewood Press, Edgewood, Maryland, USA, 1972.
*Gladioli for Everyone* J. Bentley Garrity, David and Charles, Newton Abbot, 1975

For reference reading material relating to species, see the bibliography in the relevant chapter; similarly for reference reading on breeding.

# Glossary

**Abort** to grow in an unnatural way — usually leading to a failure to flower.

**Anthocyanin** a colour-determining plastid (qv) responsible for imparting lavender-red colour to petals.

**Anthoxanthin** a colour plastid imparting cream or yellow.

**Basal plate** the central disc on the underside of a corm.

**Bigeneric** involving two different genera (i.e. botanical sub-divisions) as in hybrid Homoglads (*Homoglossum × Gladiolus*).

**Chorophyll** a very complicated organic chemical compound containing magnesium which imparts the green colour to parts of a plant and is vital as the agent promoting photosynthesis (qv), whereby the sun's energies and light are used to produce growth and changes in the plant.

**Chromosome** inheritance-determining chain of genes that may be detected during cell division using an electron microscope.

**Clone** all the vegetative material originating from an individual seed.

**Cultivar** (cv) a compound word derived from condensing 'cultivated variety'. This word is used to distinguish clonal stock from the naturally occurring species. A garden hybrid.

**Delphinidin** the plastid responsible for blue colouring in flowers — rarely occurs in the genus *Gladiolus*.

**Embryonic** in the very earliest stages of development.

**Filiform roots** the feeding roots that develop on the upper part of the corm, coming from the shoot.

**Fusarium** a fungal infection (*Fusarium oxysporum*) that invades growing plants and causes premature yellowing.

**Laciniated (Latin form; Lacinatus)** of petals, cut into at the edges giving a fringed effect.

**Malvidin** the chemical that produces the purple or violet colour in flowers.

**Nectar** a sweet, sugary syrup liquid produced deep in the throat of some gladioli to attract pollinating insects.

**Photosynthesis** a chemical reaction induced by ultra-violet light in the presence of chorophyll creating energy to sustain the plant.

**Polyploid** having numerous sets of chromosomes number unspecified but in excess of four (4).

**Plastid** a group of cells containing colour-inducing pigments that give specific colours to petals on a flower.

**Septoria** a fungus disease that attacks corms causing 'hard rot'.

**Spores** micro-organisms spread by low-development plants like ferns, or by fungus. Equivalent to seeds in higher-form plants.

**Sport** a natural mutation — usually a colour change within a cultivar stock — not brought about by hybridising.

**Stolon** the feeding tube from corm to cormlet.

**Stromatina** the fungus that causes 'neck rot'.

**Thrips** the common name for *Taeniothrips simplex*, a minute black and white flying insect that attacks gladioli.

# General Index